UCLA Symposia on Molecular and Cellular Biology, New

Series Editor, C. Fred Fox

Please contact the publisher for information about previous titles in thi

Tissue Engineering

Tissue Engineering

Proceedings of a Workshop Held at Granlibakken,
Lake Tahoe, California, February 26–29, 1988

Editors

Richard Skalak
Bioengineering Institute
Columbia University
New York, New York

C. Fred Fox
Department of Microbiology
University of California
at Los Angeles
Los Angeles, California

Alan R. Liss, Inc. • New York

Address all Inquiries to the Publisher
Alan R. Liss, Inc., 41 East 11th Street, New York, NY 10003

Library of Congress Cataloging-in-Publication Data

Tissue engineering.

(UCLA symposia on molecular and cellular biology ;
new ser., v. 107)
   Includes bibliographies and index.
   1. Tissues—Congresses.   2. Cell physiology—
Congresses.   3. Biomedical engineering—Congresses.
I. Skalak, Richard.   II. Fox, C. Fred.   III. Fung,
Y.C. (Yuan-cheng), 1919-   .   IV. Series.
[DNLM: 1. Biocompatible Materials—congresses.
2. Biomechanics—congresses.   3. Biomedical Engineering
—congresses.   4. Blood Vessel Prosthesis—congresses.
5. Implants, Artificial—congresses.   6. Muscles—physiology—congresses.   W3 U17N new ser. v.107 /
WE 500 T616 1988]

QP88.T58   1988   610'.28   88-26675
ISBN 0-8451-4706-4

Contents

VI. RESTORATION AND MAINTENANCE OF NEUROLOGICAL FUNCTION

A. MOTOR AND CENTRAL NERVE REGROWTH

B. FUNCTIONS OF CELLS AND TISSUE IMPLANTS

VII. THE HEMATOPOIETIC SYSTEM

Contributors

A. Adimora, Department of Medicine and Biochemistry, Montefiore Medical Center, Albert Einstein College of Medicine, Bronx, NY 10467; present address: Department of Medicine, Harlem Hospital, New York, NY 10037 **[109]**

P. Aebischer, Artificial Organs Laboratory, Brown University, Providence, RI 02912 **[211,257]**

Harold Alexander, Department of Bioengineering, Hospital for Joint Diseases Orthopaedic Institute, New York, NY 10003 **[189]**

Holly K. Ault, Department of Mechanical Engineering, Worcester Polytechnic Institute, Worcester, MA 01609 **[167]**

V. Bengualid, Department of Medicine and Biochemistry, Montefiore Medical Center, Albert Einstein College of Medicine, Bronx, NY 10467 **[109]**

Merton R. Bernfield, Department of Pediatrics, Stanford University Medical Center, Stanford, CA 94305 **[69]**

Haim I. Bicher, Valley Cancer Institute, Panorama City, Los Angeles, CA 91402 **[145]**

Lars M. Bjursten, Department of Anatomy, University of Göteborg, S-40033 Göteborg, Sweden **[115]**

Jonathan Black, Department of Orthopaedic Surgery, School of Medicine, and Department of Bioengineering, School of Engineering and Applied Science, University of Pennsylvania, Philadelphia, PA 19104-6081 **[195,207]**

E.A. Blumberg, Department of Medicine and Biochemistry, Montefiore Medical Center, Albert Einstein College of Medicine, Bronx, NY 10467; present address: Department of Medicine, Hahnemann University, Philadelphia, PA 19102 **[109]**

Norman C. Blumenthal, Department of Bioengineering, Hospital for Joint Diseases Orthopaedic Institute, New York, NY 10003 **[189]**

George Boder, Department of Molecular and Cell Biology, Lilly Research Labs, Indianapolis, IN 46285 **[209]**

Karen Borg, Department of Pathology, University of South Carolina, Columbia, SC 29208 **[51]**

The numbers in brackets are the opening page numbers of the contributors' articles.

Thomas K. Borg, Department of Pathology, University of South Carolina, Columbia, SC 29208 **[51]**

Steven Boyce, Department of Surgery, University of California at San Diego Medical Center, San Diego, CA 92103 **[81]**

Carl T. Brighton, Department of Orthopaedic Surgery, School of Medicine, and Department of Bioengineering, School of Engineering and Applied Science, University of Pennsylvania, Philadelphia, PA 19104-6081 **[195]**

Anthony Carabasi, Department of Surgery, Thomas Jefferson University, Philadelphia, PA 19107 **[11]**

Dennis R. Carter, Department of Mechanical Engineering, Stanford University, Stanford, CA 94305; and Rehabilitation Research and Development Center, VA Medical Center, Palo Alto, CA 94304 **[173]**

Thomas Carter, Department of Surgery, Thomas Jefferson University, Philadelphia, PA 19107 **[11]**

David T. Cheung, Laboratory of Connective Tissue Biochemistry, Departments of Biochemistry and Medicine, University of Southern California School of Medicine and Orthopaedic Hospital, Los Angeles, CA 90007 **[137]**

P. Collins, Departments of Oral Pathology and Periodontology, Dows Institute for Dental Research, College of Dentistry, University of Iowa, Iowa City, IA 52242 **[121]**

Clark K. Colton, Department of Chemical Engineering, Massachusetts Institute of Technology, Cambridge, MA 02139 **[217]**

Stephen C. Cowin, Department of Biomedical Engineering, Tulane University, New Orleans, LA 70118 **[181]**

Dennis D. Cunningham, Department of Microbiology and Molecular Genetics, College of Medicine, University of California at Irvine, Irvine, CA 92717 **[263]**

Don DiMasi, Biotechnology Engineering Center, Tufts University, Medford, MA 02155 **[299]**

Keith E. Dionne, Department of Chemical Engineering, Massachusetts Institute of Technology, Cambridge, MA 02139 **[217]**

C. Doller, Department of Neurosurgery, Columbia Presbyterian Medical Center, New York, NY 10032; and Department of Developmental Genetics, Case Western Reserve University, Cleveland, OH 44106 **[249]**

Brian Durig, Department of Mechanical Engineering, University of South Carolina, Columbia, SC 29208 **[51]**

Lars E. Ericson, Department of Anatomy, University of Göteborg, S-40033 Göteborg, Sweden **[115]**

Jerome L. Finkelstein, Burn Center, New York Hospital-Cornell Medical Center, New York, NY 10021 **[73]**

David M. Flynn, Mechanical Engineering, Worcester Polytechnic Institute, Worcester, MA 01609 **[167]**

Tanya Foreman, Department of Surgery, University of California at San Diego Medical Center, San Diego, CA 92013 **[81]**

C. Fred Fox, Department of Microbiology, University of California at Los Angeles, Los Angeles, CA 90024 **[xix, 97, 151]**

William J. Freed, Preclinical Neurosciences Section, National Institute of Mental Health, St. Elizabeth's Hospital, Washington, DC 20032 [243,273]

Y.-C. Fung, Department of AMES/ Bioengineering, University of California at San Diego, La Jolla, CA 92093 [xix,45]

P.M. Galletti, Artificial Organs Laboratory, Brown University, Providence, RI 02912 [19,211]

M.B. Goddard, Artificial Organ Laboratory, Brown University, Providence, RI 02912 [19]

Daniel A. Grande, Department of Bioengineering, Hospital for Joint Diseases Orthopaedic Institute, New York, NY 10003 [189]

Peter Grigg, Department of Physiology, University of Massachusetts Medical School, Worcester, MA 01605 [167]

Mark Grise, Biotechnology Engineering Center, Tufts University, Medford, MA 02155 [299]

Anthony G. Gristina, Section of Orthopedic Surgery Center, Wake Forest University Medical Center, Winston-Salem, NC 27103 [99]

Farshid Guilak, Orthopaedic Research Laboratory, Departments of Orthopaedic Surgery and Mechanical Engineering, Columbia University, New York, NY 10032 [161]

David Gurwitz, Department of Microbiology and Molecular Genetics, College of Medicine, University of California at Irvine, Irvine, CA 92717 [263]

Allan Haberman, Biotechnology Engineering Center, Tufts University, Medford, MA 02155 [299]

John Hansbrough, Department of Surgery, University of California at San Diego Medical Center, San Diego, CA 92103 [81]

V.B. Hatcher, Department of Medicine and Biochemistry, Montefiore Medical Center, Albert Einstein College of Medicine, Bronx, NY 10467 [109]

John M. Hefton, Burn Center, New York Hospital-Cornell Medical Center, New York, NY 10021 [73]

Allen H. Hoffman, Mechanical Engineering, Worcester Polytechnic Institute, Worcester, MA 01609 [167]

Kajsa-Mia Holgers, Department of Anatomy and Department of ENT, Sahlgrens Hospital, University of Göteborg, S-40033 Göteborg, Sweden [115]

Bruce Jacobson, Biotechnology Engineering Center, Tufts University, Medford, MA 02155 [299]

Bruce E. Jarrell, Department of Surgery, Thomas Jefferson University, Philadelphia, PA 19107 [1,11,241]

Michel Kliot, Department of Neurosurgery, Columbia Presbyterian Medical Center, New York, NY 10032; and Department of Developmental Genetics, Case Western Reserve University, Cleveland, OH 44106 [249,269]

Kevin J. Lafferty, Barbara Davis Center for Childhood Diabetes, Department of Microbiology/ Immunology and Pediatrics, University of Colorado Health Science Center, Denver, CO 80262 [231]

Ana Lages, Biotechnology Engineering Center, Tufts University, Medford, MA 02155 [299]

J.E. Lemons, Department of Biomaterials, University of Alabama at Birmingham, Birmingham, AL 35294 **[201]**

Philip Litwak, Thoratec Laboratories Corporation, Berkeley, CA 94710 **[25]**

F.D. Lowy, Department of Medicine and Biochemistry, Montefiore Medical Center, Albert Einstein College of Medicine, Bronx, NY 10467 **[109]**

Michael R. Madden, Burn Center, New York Hospital-Cornell Medical Center, New York, NY 10021 **[73]**

Bonnie Miller, Department of Pathology, University of South Carolina, Columbia, SC 29208 **[51]**

Van C. Mow, Orthopaedic Research Laboratory, Departments of Orthopaedic Surgery and Mechanical Engineering, Columbia University, New York, NY 10032 **[153,161]**

Quentin N. Myrvik, Department of Microbiology and Immunology, Wake Forest University Medical Center, Winston-Salem, NC 27103 **[99]**

Robert M. Nerem, School of Medical Engineering, Georgia Institute of Technology, Atlanta, GA 30332-0405 **[5]**

Dimitri Nicolakis, Biotechnology Engineering Center, Tufts University, Medford, MA 02155 **[299]**

Marcel E. Nimni, Laboratory of Connective Tissue Biochemistry, Departments of Biochemistry and Medicine, University of Southern California School of Medicine and Orthopaedic Hospital, Los Angeles, CA 90007 **[127,137]**

George D. Pappas, Department of Anatomy and Cell Biology, University of Illinois College of Medicine, Chicago, IL 60612 **[283,289,295]**

Pauline Park, Department of Surgery, Thomas Jefferson University, Philadelphia, PA 19107 **[11]**

D. Patel, Department of Medicine and Biochemistry, Montefiore Medical Center, Albert Einstein College of Medicine, Bronx, NY 10467 **[109]**

Walter Peters, Department of Mechanical Engineering, University of South Carolina, Columbia, SC 29208 **[51]**

Robert A. Peura, Biomedical Engineering Program, Worcester Polytechnic Institute, Worcester, MA 01609 **[223]**

Solomon R. Pollack, Department of Orthopaedic Surgery, School of Medicine, and Department of Bioengineering, School of Engineering and Applied Science, University of Pennsylvania, Philadelphia, PA 19104-6081 **[195]**

Anthony Ratcliffe, Orthopaedic Research Laboratory, Departments of Orthopaedic Surgery and Mechanical Engineering, Columbia University, New York, NY 10032 **[161]**

Karel Rakusan, Department of Physiology, University of Ottawa, Ottawa, Ontario K1H 8M5, Canada **[57]**

John L. Ricci, Department of Bioengineering, Hospital for Joint Diseases Orthopaedic Institute, New York, NY 10003 **[189]**

Peter D. Richardson, Division of Engineering, Brown University, Providence, RI 02912 **[39,65]**

A. Jean Robinson, Thoratec Laboratories Corporation, Berkeley, CA 94710 **[25]**

A.W. Romanowski, Departments of Oral Pathology and Periodontology, Dows Institute for Dental Research, College of Dentistry, University of Iowa, Iowa City, IA 52242 **[121]**

Deborah Rose, Department of Surgery, Thomas Jefferson University, Philadelphia, PA 19107 **[11]**

James D. Russell, Division of Biomedical Sciences, School of Graduate Studies, and Department of Biochemistry, School of Medicine, Meharry Medical College, Nashville, TN 37208 **[87]**

Shirley B. Russell, Division of Biomedical Sciences, School of Graduate Studies, and Department of Biochemistry, School of Medicine, Meharry Medical College, Nashville, TN 37208 **[17,87]**

Jacqueline Sagen, Department of Anatomy and Cell Biology, University of Illinois, Chicago, IL 60612 **[283, 289]**

Jerome S. Schultz, Center for Biotechnology and Bioengineering, University of Pittsburgh, Pittsburgh, PA 15260 **[313]**

Saul S. Schwarz, Department of Neurosurgery, Naval Hospital Bethesda, and National Institute of Mental Health, St. Elizabeth's Hospital, Washington, DC 20032 **[273]**

J. Siegal, Department of Neurosurgery, Columbia Presbyterian Medical Center, New York, NY 10032; and Department of Developmental Genetics, Case Western Reserve University, Cleveland, OH 44106 **[249]**

J. Silver, Department of Neurosurgery, Columbia Presbyterian Medical Center, New York, NY 10032; and Department of Developmental Genetics, Case Western Reserve University, Cleveland, OH 44106 **[249]**

Richard Skalak, Columbia University, Bioengineering Institute, New York, NY 10027 **[xix,321]**

G.M. Smith, Department of Neurosurgery, Columbia Presbyterian Medical Center, New York, NY 10032; and Department of Developmental Genetics, Case Western Reserve University, Cleveland, OH 44106 **[249]**

G. Soldani, Artificial Organ Laboratory, Brown University, Providence, RI 02912 **[19]**

Carol A. Spatz, Thoratec Laboratories Corporation, Berkeley, CA 94710 **[25]**

Katherine H. Sprugel, Department of Pathology, University of Washington, Seattle, WA 98195 **[93]**

Christopher A. Squier, Departments of Oral Pathology and Periodontology, Dows Institute for Dental Research, College of Dentistry, University of Iowa, Iowa City, IA 52242 **[121,151,319]**

Lisa Staiano-Coico, Burn Center, New York Hospital-Cornell Medical Center, New York, NY 10021 **[73]**

Randall W. Swartz, Biotechnology Engineering Center, Tufts University, Medford, MA 02155 **[151, 297, 299]**

Louis Terracio, Department of Anatomy, University of South Carolina, Columbia, SC 29208 **[51]**

Peter Thomsen, Department of Anatomy, University of Göteborg, S-40033 Göteborg, Sweden **[115]**

Anders Tjellström, Department of Anatomy, University of Göteborg, S-40033 Göteborg, Sweden; present address: Department of ENT, Sahlgrens Hospital, University of Göteborg, S-40033 Göteborg, Sweden **[115]**

D.C. Tompkins, Department of Medicine and Biochemistry, Montefiore Medical Center, Albert Einstein College of Medicine, Bronx, NY 10467 **[109]**

Joel S. Trupin, Division of Biomedical Sciences, School of Graduate Studies, and Department of Biochemistry, School of Medicine, Meharry Medical College, Nashville, TN 37208 **[87]**

George Truskey, Biotechnology Engineering Center, Tufts University, Medford, MA 02155 **[299]**

Vincent T. Turitto, Department of Medicine, Mount Sinai Medical Center, New York, NY 10029 **[31,37]**

S. Tyrrell, Department of Neurosurgery, Columbia Presbyterian Medical Center, New York, NY 10032; and Department of Developmental Genetics, Case Western Reserve University, Cleveland, OH 44106 **[249]**

Yi Wang, Barbara Davis Center for Childhood Diabetes, Department of Microbiology/Immunology and Pediatrics, University of Colorado Health Science Center, Denver, CO 80262 **[231]**

Robert S. Ward, Thoratec Laboratories Corporation, Berkeley, CA 94710 **[25]**

Jennifer S. Wayne, Orthopaedic Bioengineering Laboratory, University of California at San Diego, and San Diego VA Medical Center, La Jolla, CA 92093 **[155]**

Lawrence X. Webb, Section of Orthopedic Surgery, Wake Forest University Medical Center, Winston-Salem, NC 27103 **[99]**

Stuart Williams, Department of Surgery, Thomas Jefferson University, Philadelphia, PA 19107 **[11]**

Savio L-Y. Woo, Orthopaedic Bioengineering Laboratory, University of Califorina at San Diego, and San Diego VA Medical Center, La Jolla, CA 92093 **[155]**

Martin L. Yarmush, Department of Chemical Engineering, Massachusetts Institute of Technology, Cambridge, MA 02139 **[217]**

Iskender Yilgor, Thoratec Laboratories Corporation, Berkeley, CA 94710 **[25]**

Preface

This volume is a record of papers and discussions presented at a workshop on "Tissue Engineering" held at Granlibakken, Lake Tahoe, California, February 26-29, 1988. The workshop was assembled at the recommendation of a Panel on Tissue Engineering which met at the National Science Foundation on October 28, 1987. This Panel consisted of representatives from the NSF, NIH, ONR, DOE, NASA, Red Cross, and university representatives from areas of cell biology, medicine and bioengineering. On October 29-30, 1987, the Division of Emerging Technology of NSF held a Forum on Issues, Expectations and Prospects for the Emerging Technology Research Initiations in Washington, D.C. At this Forum, Tissue Engineering was recommended for development as a vital, emerging technology for rapid industrial exploitation, and the idea that a workshop be held to examine the issues was given substantial support. It proved to be relatively easy to assemble a knowledgeable group of investigators engaged in research related to Tissue Engineering who were eager to share their views and discuss future prospects. Approximately seventy-five people attended the workshop; these included university, governmental agency, and industrial representatives. The papers and discussions included in this volume indicate the wide range of tissues, procedures and background basic science that was represented. It is our hope that this volume may transmit some of the excitement and rapport which was developed at the workshop over a wide spectrum of subjects, disciplines and backgrounds of the participants.

For the purposes of the workshop and further discussion, the following definition was developed:

"Tissue Engineering" is the application of principles and methods of engineering and life sciences toward fundamental understanding of structure-function relationships in normal and pathological mammalian tissues and the development of biological substitutes to restore, maintain, or improve tissue functions.

The basic point of the above definition is that tissue engineering involves the use of living cells or living cells plus their extracellular products in development of biological substitutes for replacements as opposed to the use of inert implants. The definition is intended to encompass procedures in which the replacements may consist of cells in suspension, cells implanted on a scaffold such as collagen and cases in which the replacement consists entirely of cells and their extracellular products.

It was clear from the workshop that this type of biological replacement is an area of current excitement in several medical areas which, however, have some common basis and common problems in establishing practical techniques. Further, many different disciplines and effects are involved. These may be summarized in short as chemical, mechanical and electrical. As may be seen from this proceedings volume, these different stimuli may all be cogent and useful in the development of tissue engineering. To be more specific, the main discussions at the workshop included:

Vascular and Endothelial Cell Technology
Skin and Connective Tissue
Musculoskeletal System and Orthopaedic Surgery
The Nervous System
Islet and Chromaffin Cells as Transplants
The Hematopoietic System
Mathematical Modeling.

The workshop was especially successful because a wide range of senior medical, biological and engineering scientists expressed interest in seeking commonality of principles and procedures, and shared their experiences and views. It is clear that the workshop participants consider "Tissue Engineering" as both a forefront of biomedical engineering research and an area that has broad and promising practical applications for improving the quality of life.

The workshop was highly successful at a personal level in developing a dialogue between diverse groups of bioengineers, biological scientists and medical researchers. It is the hope of the organizers in producing this volume that it may stimulate the further development of this rapport. In any case, the workshop has clearly identified the utility and promise of cultivation of Tissue Engineering as a multidisciplinary field and an emerging technology.

Richard Skalak
C. Fred Fox
Bert Fung

Tissue Engineering, pages 1–3
© 1988 Alan R. Liss, Inc.

I. Vascular and Endothelial Cell Technology

- *Microcirculation: Endothelial Cells, Angiogenesis*

- *Vascular Prosthesis*

- *Blood Vessel, Soft Tissue Mechanics and Remodeling*

Vascular surgery is an area of medicine that has developed greatly over the last thirty years. The number of vascular reconstructions has risen tremendously during that period of time. It has been estimated, for example, in 1982 that over 340,000 major vascular procedures were performed. Over half of these procedures are peripheral reconstructions and generally involve the use of some type of graft as a conduit to bypass an occluded arterial segment. For the lower extremity, this frequently involves the use of the saphenous vein as a bypass conduit. This has been very successful with 90% patency rates long term. Many patients do not have such a vessel and therefore require the use of a synthetic graft for bypass. The results using synthetic grafts have been directly proportional to the length and diameter of the graft. As the graft becomes longer and smaller in diameter, the results drop dramatically with only a small fraction of these grafts being patent at the end of one year. Coronary artery bypass grafting is also a very active area in arterial surgery. Over 100,000 coronary bypass grafts are done per year and an equal number of coronary artery angioplasties are performed. Again, native vessel including saphenous vein or internal mammary artery is the graft of choice and both have shown excellent long-term patency rates. The results when synthetic grafts are used in this location have been poor.

The major cause of limitation in these grafts, both in the coronary artery and peripheral circulation is a complex issue. Two events appear to be firmly established. First, cellular hyperplasia frequently occurs at or around the anastomosis between the vascular graft and the native artery. This anastomotic hyperplasia frequently results in progressive stenosis of the anastomosis resulting in eventual occlusion and failure of the graft.

The second issue relates to the failure of the naturally occurring endothelium present on normal vessels to occur spontaneously on prosthetic surfaces. Thus, the new surface that is in contact with the blood is somewhat thrombogenic.

Cellular biology and molecular biology approaches to the anastomotic hyperplasia problem are likely to provide information of value in designing therapeutic interventions. Growth factors or cell growth inhibitors may be useful in this regard. Study of the thrombogenicity of a surface has revolved around the use of biopolymers as well as attempting to reestablish endothelial cells upon this surface. It is reasonable that reestablishment of endothelium on such a surface would impart a certain degree of antithrombogenicity to that surface.

The construction of blood-contacting surfaces has been approached through three major experimental directions which include construction of non-thrombotic polymers, resorbable polymers and cell-lined polymers. The search for a non-thrombotic polymer has failed to produce clinically useful products over recent years.Several new possibilities are on the horizon. Future success depends upon both novel chemistry applications as well as gaining an understanding of the influence of mechano-physical properties on blood coagulation. A major limitation is our incomplete knowledge of the coagulation system and its multitude of reactions with different surfaces.

Another area of importance is development of resorbable polymers and their interactions with the cellular milieu. Healing and the development of normal structures typical of the arterial wall are events that should be understood and engineered to produce optimal results. Although there is concern for mechanical stability of these polymers in the arterial tree, a strong potential benefit is the absence of a chronic foreign body and thus resistance to infection.

Cell-lined polymers capable of interacting actively with blood represent a third approach to a vascular conduit. Although many different cell types may be capable of assuming this role, most efforts have been directed toward endothelial cell coverage of polymers. There are many issues in this area. The use of endothelial cell growth factors to spontaneously endothelialize a surface is exciting. Many interdisciplinary questions arise with respect to immobilization and controlled release of growth factors from a substrate and subsequent cell migration and behavior on the surface. A second long term issue relates to whether cells that have initially been stimulated with growth factors are capable of returning to basel metabolic and reproductive functions once the surface has been re-endothelialized.

Endothelial cell behavior under different conditions is an important area where biologists and engineers actively interact. Endothelial cells are known to be strongly influenced by environmental, physical and biochemical perturbations at both the luminal and abluminal surfaces. Blood-factors are known to produce pro- and anticoagulant functions as well as interactions with bacteria. Shear stress conditions affect cell structure and function. Cell substrate conditions present on both prosthetic and naturally-occurring surfaces of extracellular matrix are known to affect widely diverse properties including cell function, differentiation, morphology and attachment properties. Cell attachment occurs to many surfaces over a time scale of minutes. Manipulation of these variables involves application of many engineering principles and offers the possibility of rapidly endothelializing a surface. Once endothelialized, the variables that affect function can be examined. Optimization of all these factors to produce a natural vesel replacement is a complex bioengineering project.

Tissue Engineering, pages 5–10

ENDOTHELIAL CELL RESPONSES TO SHEAR STRESS: IMPLICATIONS IN THE DEVELOPMENT OF ENDOTHELIALIZED SYNTHETIC VASCULAR GRAFTS[1]

Robert M. Nerem

School of Mechanical Engineering
Georgia Institute of Technology
Atlanta, GA 30332-0405

ABSTRACT   Vascular endothelial cells demonstrate a wide variety of responses to fluid dynamic shear stress. This includes decreased cell proliferation in sub-confluent monolayers exposed to flow. Recent studies have established that endothelial pre-seeding of small diameter, synthetic vascular grafts favorably influences graft patency. In order to further enhance graft patency, consideration of the influence of the fluid dynamic environment in the determination of optimal conditions for achieving and maintaining endothelial monolayer integrity may be important.

INTRODUCTION

In vascular bypass surgery there is a need for small diameter (4-6 mm) synthetic grafts, which will remain patent for long periods of time. The ideal synthetic prosthetic vascular graft will possess thrombo-resistant properties and will not be subject to narrowing, as produced by either atherosclerotic plaque development or smooth muscle cell proliferation and associated connective tissue synthesis.

A number of laboratories have been experimenting with the use of endothelial pre-seeding of synthetic vascular

[1]This work was supported by National Science Foundation Grant EET-8796223 and National Institutes of Health Grant HL-26890

grafts. These studies have established that endothelial pre-seeding does influence graft patency favorably (1). However, insufficient consideration has been given to the fluid dynamic environment of such grafts. This is in spite of the considerable work that has been carried out over the past 10 years on the influence of shear stress on endothelial cells (EC).

THE IN VIVO FLUID DYNAMIC ENVIRONMENT

A major hypothesis underlying our work is that (i) the structure and function of vascular endothelial cells are determined by the environment in which they reside and (ii) this environment is not only biochemical in nature, but also mechanical. In vivo the fluid dynamic environment may be characterized as in general laminar and unsteady with highly complex flow patterns due to the branching and curvature of the blood vessels. This somewhat tortuous geometry is highly individualistic, i.e., its detailed features vary from one individual to the next - sometimes considerably, and it has been suggested in the literature that one of the risk factors for atherosclerosis may be a person's individual geometry. Whether this is the case or not, it is clear that the specific details of the in vivo fluid dynamic environment will depend on an individual's exact geometry, whether or not the individual is in a resting condition or exercising, and the specific artery of interest.

The left coronary vascular system will be used as an example, and for a human the diameter of the larger left coronary arteries is in the range of 3-4 mm (2). Using a resting flow rate of 100 ml/min, the mean Reynolds number will be in the range of 100-200, for a normal heart rate the Womersley parameter, $\alpha = R(\omega/\nu)^{1/2}$, will be of order one, and the mean wall shear stress will be in the range of 10-15 dynes/cm^2 (2). However, values of wall shear stress may easily increase significantly due to velocity profile skewing in regions of branching, e.g., on flow dividers, on a transient basis during the cardiac cycle as the flow pulses, and for conditions of strenuous exercise. In some cases, peak shear stress values may easily exceed 100 dynes/cm^2.

VASCULAR ENDOTHELIAL CELL RESPONSES TO SHEAR STRESS

Below the level of shear stress required for endothelial denudation, it appears that in vivo EC geometry and orientation are influenced by hemodynamics. Other studies have demonstrated a relationship of macromolecule uptake and cell turnover (3) with endothelial cell morphology, thus suggesting implicitly a relationship to the shear stress environment.

A number of groups have initiated studies on the influence of shear stress on cultured EC. Dewey, et al. (4) found that subconfluent endothelial cultures, continuously exposed to 1-5 dynes/cm^2, proliferated at a rate comparable to that of static cultures and reached the same saturation density ($\sim 10^5$ cells/cm^2). For a confluent EC monolayer exposed to a laminar shear stress, cell shape becomes more elongated and oriented with the direction of flow and a coincident cytoskeletal reorganization occurs (4-6).

Other EC functions also have been shown to be influenced by fluid-imposed shear stress. Horseradish peroxidase (HRP) endocytosis results (7) indicate that, although HRP endocytosis was unchanged by either steady shear stress or a 1 Hz modulation of shear stress, it was responsive to step-changes of durations on the order of 10 minutes or more. Davies, et al. (8), in the only study to investigate the effect of turbulence, found enhanced cell proliferation for a confluent EC monolayer exposed to a turbulent shear stress. This is in contrast to a confluent monolayer exposed to a laminar shear stress, where there is no enhanced proliferation.

In our own laboratory, we have studied the response of cultured bovine aortic endothelial cells to steady shear stresses up to 100 dynes/cm^2. These studies have demonstrated that the elongation and orientation of EC in response to shear stress is dependent not only on the level of shear stress and the duration of exposure, but also on the substrate to which the cells are attached (5). Furthermore, there is a major reorganization of the cytoskeleton in response to shear stress which reflects itself in increased cell stiffness (6).

For subconfluent BAEC monolayers, we have observed an influence of shear stress on cell proliferation, as determined by measurements of the time history of cell density and by ^{3}H-thymidine incorporation and autoradiography. This effect of shear stress is further enhanced by the presence of flow pulsatility as are some other characteristics of vascular EC.

IMPLICATIONS FOR ENDOTHELIALIZED VASCULAR GRAFTS

A number of studies have established that endothelial pre-seeding of small diameter (4 mm) synthetic vascular polytetrafluoroethylene (PTFE) or dacron grafts favorably influences graft patency (1). EC-seeded grafts exhibit less thrombosis and also less narrowing. At least part of the graft narrowing is due to smooth muscle cell proliferation and connective tissue synthesis at the anastomotic sites (9). Studies using either total platelet counts or radiolabeled platelets have reported that pre-seeded grafts are associated with less platelet utilization, as reflected in longer platelet survival time (10), consistent with the view that graft endothelialization confers thrombo-resistance. Further, there is now evidence that the use of homologous as distinct from autologous endothelium for pre-seeding has no discernible disadvantage (11).

These studies have established the efficacy of partial pre-endothelialization of synthetic vascular grafts. However, little consideration has been given to the possible role of flow factors, and the fluid dynamic environment in which the cells reside may be all important, e.g., for cell attachment and cell proliferation.

Furthermore, the fluid dynamic environment may be important to other questions relative to graft patency. For example, blood mononuclear phagocytes play a significant role in the pathogenesis of atherosclerosis and are pivotal in the inflammatory responses of the arterial wall. Their roles and significance in determining the fate of pre-endothelialized synthetic grafts remains unknown, as do the factors which might regulate their recruitment and attachment to the endothelialized graft. A spectrum of secretory products of the mononuclear phagocyte could also influence graft patency. It seems important, therefore, to consider the influence of shear stress on the recruitment and attachment of blood monocytes to the endothelium.

DISCUSSION

Although previous studies have suggested that endothelial pre-seeding of synthetic vascular grafts enhances graft patency, there are questions still to be answered. These include what source will be used for human EC and is it really true that homologous EC are at no distinct disadvantage to autologous EC?

To this we now add questions relating to the influence of the fluid dynamic environment and in particular shear stress. In addition to influences on EC biology, there are also such questions as should the pre-seeded endothelium on the graft surface be grown to confluency prior to implantation? If so, should this be done in static culture or in the presence of flow? Once grown to confluency, should the endothelialized graft be fluid-dynamically pre-conditioned? If so, what is the optimal flow environment that should be used? Can it be a steady flow or is it necessary to duplicate the arterial flow waveform? What level of shear stress should be employed? Should the flow be laminar or turbulent? These types of questions not only motivate our interest, but are important to address in the development of pre-seeded endothelialized synthetic vascular grafts.

ACKNOWLEDGMENTS

The author acknowledges the many contributions provided by Dr. Murina J. Levesque to the cell culture research in our laboratory and the several discussions with Drs. Colin J. Schwartz and Eugene A. Sprague on the role of flow factors in endothelial-seeded vascular grafts.

REFERENCES

1.	Stanley JC, Burkel WE, Graham LM, Lindbald B (1985). Endothelial cell seeding of synthetic vascular prostheses. Acta Chir Scand Suppl. 529:17.

2.	Nerem RM, Seed WA (1983). Coronary artery geometry and its fluid mechanical implications. In Schettler G, et al.: "Fluid Dynamics as a Localizing Factor for Atherosclerosis," Berlin Heidelberg: Springer Verlag, p. 51.

3. Schwartz CJ, Sprague EA, Fowler SR, Kelley JL (1983). Cellular participation in atherogenesis: selected facets of endothelium, smooth muscle and peripheral blood monocyte. In Schettler G, et al.: "Fluid Dynamics as a Localizing Factor for Atherosclerosis," Berlin Heidelberg: Springer Verlag, p. 200.

4. Dewey CF, Jr., Bussolari SR, Gimbrone MA, Jr., Davies PF (1981). The dynamic response of vascular endothelial cells to fluid shear stress. ASME J Biomech Engr 103:177.

5. Levesque MJ, Nerem RM (1985). The elongation and orientation of cultured endothelial cells in response to shear stress. ASME J Biomech Engr 107:341.

6. Sato M, Levesque MJ, Nerem RM (1987). Micropipette aspiration of cultured bovine aortic endothelial cells exposed to shear stress. Arteriosclerosis 7:276.

7. Dewey CF, Jr. (1984). Effects of fluid flow on living vascular cells. ASME J Biomech Engr 106:31.

8. Davies PF, Remuzzi A, Gordon EF, Dewey CF, Jr., Gimbrone MA, Jr. (1986). Turbulent fluid shear stress induces vascular endothelial cell turnover in vitro. Proc Natl Acad Sci 2114.

9. Clowes AW, Gown AM, Hanson SR, Reidy MA (1985). Mechanisms of arterial graft failure. I. Role of cellular profileration in early healing of PTFE prostheses. Am J Path 118:43.

10. Clagett GP, Burkel WE, Sharefkin JB, Ford JW, Hufnagel H, Vinter DW, Kahn RH, Graham LM, Stanley JC, Ramwell PW (1984). Antithrombotic character of canine endothelial cell-seeded arterial prostheses. Surg Forum 33:471.

11. Zamora JL, Navarro LT, Ives CL, Weilbaecher DG, Gao ZR, Noon GP (1986). Seeding of arteriovenous prostheses with homologous endothelium. J Vasc Surg. 3:860.

Tissue Engineering, pages 11–15
© 1988 Alan R. Liss, Inc.

HUMAN ENDOTHELIAL CELL INTERACTIONS WITH VASCULAR GRAFTS[1]

Bruce Jarrell[2], Stuart Williams, Pauline Park,
Thomas Carter, Anthony Carabasi, Deborah Rose

Department of Surgery, Thomas Jefferson University
Philadelphia, PA 19107

ABSTRACT Endothelial cell (EC) lining may restore
biological function to blood contacting surfaces.
Essential components include EC; rapid, firm
adherence; and rapid resumption of function. Using
human large microvessel EC, attachment proceeds
rapidly and is capable of producing a spread cell
monolayer within one hour. Microscopy suggests that
EC attachment involves the extension of a foot
process to the surface prior to attachment and
spreading. In our shear stress detachment assay,
strong attachment forces capable of withstanding
aortic shear stresses develop over minutes as cells
undergo the process of spreading. The force of
attachment varies with function of the attachment
surface, the type of EC and may even vary between
nuclear and cytoplasmic portion of the cell. Firmly
attached EC monolayers can be produced rapidly on
prosthetic and natural vessels suggesting use in
both vessel replacement and vessel repair. When
implanted into animals, these endothelialized
surfaces exhibit intact monolayers and
nonthrombogenicity for periods in excess of 30 days,
indicating that in-vitro optimization of
conditions can be very useful to reflect eary
in-vivo events.

[1] This work was supported by NIH Grant
1-RO1-HL38103-01A1 and the W.W. Smith Charitable Trust
[2] Present address: Dept. of Surgery, Thomas
Jefferson University, 1025 Walnut Street, Philadelphia,
PA 19107

INTRODUCTION

Human endothelial cells (EC) form the blood-contacting surface in the body. These cells possess both well-defined and undefined mechanisms that allow interaction with the blood elements and the vessel wall. Presumably through the process of evolution, the cells have fine tuned these mechanisms to maintain homeostasis during variations in physiological conditions. In contrast, "non-living" polymer surfaces in contact with blood may possess a very limited ability to dynamically interact with blood, thus making the flexibility in a given situation much less than in a living cell. It seems logical, therefore, that prosthetic vascular-contacting surfaces should be designed to integrate the best of the polymer characteristics with the best blood contacting surface, the endothelial cell. To accomplish this goal, one must first successfully establish these cells upon a polymer surface and demonstrate "normal" morphology. Once that has been accomplished, normal function can then be established.

ENDOTHELIAL CELL ATTACHMENT PROCESS

We have approached this problem by examining the process of EC attachment to a variety of surfaces. The essential components of EC attachment have been observed on both naturally occurring surfaces such as human basement membrane (amnion) and prosthetic polymeric surfaces including polystyrene and dacron (polyethylene-terephthalate)[1,2]. The initial step is the approach of cells to the surface due to the force of gravity. This may require only several minutes to deliver a large number of cells to the surface. The next step involves the extension of an EC foot process to the surface. If the interaction between EC and surface is correct, the EC will rapidly broaden the point of attachment to the surface. EC in this state may be in a rounded morphology but still capable of resisting significant detachment forces such as seen in the abdominal aorta. The development of this strong attachment force occurs very rapidly following incubation for the majority of EC. On fibronectin-coated polystyrene, for example, strong forces are present as early as 5 minutes following incubation. Following attachment, the next step is spreading. This is a more variable process than the earlier steps and can occur at a rate ranging in

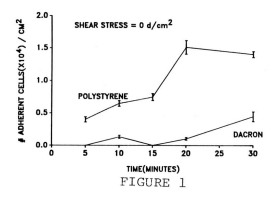

FIGURE 1

time from less than 60 minutes on an optimal surface to extended periods of time on poorer surfaces (see Figure 1). If enough EC are present on a good surface such as basement membrane or fibronectin coated polymers, complete coverage of the surface with a monolayer of EC has been observed to occur in less than one hour following incubation[3]. In an attempt to quantify the short-term mechanical stability of such a monolayer, we have developed an EC detachment assay using a rotating disc device that exposes EC to a defined shear stress[2]. In that assay, we have noted under many different experimental conditions that attached EC do not detach when exposed to shear forces that are equivalent to that observed in the aorta (see Figure 2).

CLINICAL IMPLICATIONS

The general implications of these observations relate to the establishment of a cell lining on a polymeric surface. Given a receptive surface, EC attachment proceeds rapidly and is capable of producing a detachment-resistant cell-covered surface within minutes. The EC do not necessarily have to be fully spread to resist detachment although a spread cell morphology is most likely more desirable from a functional standpoint. In addition, these observations suggest that polymers can be designed that will expedite cell attachment and spreading. These polymers would permit rapid

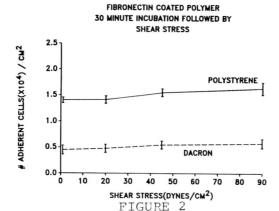

FIGURE 2

endothelialization compatible with an operating room time frame, and thus avoid the use of prolonged cell culture techniques and other complicated procedures.

While the ideal polymer for endothelialization does not exist at this time, we have tested the hypothesis of rapid endothelialization in animals using currently available vascular grafts. Using the canine model, autologous microvessel EC were isolated from fat tissue and immediately incubated with a plasma-coated dacron graft for 2 hours[4]. This is a process we term sodding. Graft were implanted as a carotid artery interposition graft. At implantation, the lumenal surface of the graft exhibited attached cells with both a spread and focally attached morphology. Following 35 days of implantation, the grafts continue to exhibit an EC monolayer with no attached fibrin, platelets or white blood cells. In contrast, control grafts demonstrated a layer of platelets, white cells and fibrin. Results of these canine studies must be analyzed in light of the fact that a suboptimal graft polymer was used. In addition, it is not certain that canine results accurately reflect endothelial cell sodded graft performance in humans. It does demonstrate, however, that a rapid attachment process is capable of producing an excellent appearing flow surface.

SUMMARY

In summary, human EC possess the ability to rapidly attach to polymeric surfaces and resist detachment. EC properties observed in in-vitro studies appear to reflect in-vivo events with respect to the generation and short-term maintenance of a monolayer. Optimization of this process may produce excellent surfaces even at the time of implantation.

REFERENCES

1. Jarrell, B.E., Williams, S.K., Solomon, L., et. al., (1986) Use of an endothelial monolayer on a vascular graft prior to implantation. Temporal dynamics with the operating room. Ann. Surg. 203 (6) 671-678.
2. Pratt, K.J., Jarrell, B.E., Williams, S.K., et. al., (1988) Kinetics of endothelial cell-surface attachment forces. In press, J. Vasc. Surg.
3. Jarrell, B.E., Williams, S.K., Pratt, et. al., (1988) Cell attachment forces regulating the immediate establishment of endothelial cell monolayers. In: Endothelialization of Vascular Grafts. P. Zilla, R. Fasol (Eds), Karger Publishing Co., Basel, Switzerland.
4. Jarrell, B.E., Williams, S.K., Stokes, G., et. al., (1986). Use of freshly isolated capillary endothelial cells for the immediate establishment of a monolayer on a vascular graft at surgery. Surgery, 100:392-399.

Tissue Engineering, pages 17–18
© 1988 Alan R. Liss, Inc.

Summary/Discussion

Microcirculation: Endothelial Cells, Angiogenesis

Shirley B. Russell

Peptide factors that promote cell recruitment and proliferation have great potential in the newly emerging field of tissue engineering, largely because they can rapidly promote the production of tissue of host origin. The use of implants containing endothelial cell growth factor to promote production of blood vessels generated discussion concerning the stability of the new vessels, identification of cells recruited to the implant, and the contribution of host genes. When implants are removed or absorbed, the new vessels disappear. Consequently, inert materials are being sought that will give permanence to the implant and stability to the vessels. The small amount of growth factor required suggests that helper genes, possibly for homologous factors such as basic fibroblast growth factor or the human stomach tumor oncogene, *hst,* are being turned on. One worthwhile goal is to identify and clone relevant gene products that are expressed in the host. Methods are also being sought to induce elastin synthesis at specific anatomical sites to promote further differentiation of the new vessels. Several cell types are recruited to the implant; however, with the exception of endothelial cells, these cell types have not yet been identified. Their identification should help to define factors that promote and regulate angiogenesis.

Synthetic vascular material seeded with endothelial cells may provide a method for producing replacement vessels with low thrombogenicity. Discussion of this approach focused on the need for better methods to assess requirements for cell attachment and proliferation *in vivo* as well as functional tests. There was a perceived need to identify and test *in vitro* synthetic materials that were similar to natural substrata for endothelial cells and to examine the requirements of endothelial cells derived from different types of blood vessels. Questions were raised about the adequacy of current methods for testing anti-thrombogenicity, adherence and deformability of cells. Mechanical effects such as shear and pressure clearly play a role in cell function. It was suggested that these effects should be measured in conjunction with assays for biochemical regulators to determine the identity and characteristics of molecules mediating these responses.

Tissue Engineering, pages 19–24
© 1988 Alan R. Liss, Inc.

TRANSIENT BIOCOMPATIBLE SCAFFOLDS FOR REGENERATION
OF THE ARTERIAL WALL

P. M. Galletti, M.B. Goddard, and G. Soldani

Artificial Organ Laboratory, Brown University

ABSTRACT The elements of a normal arterial wall may
be reorganized through natural repair processes if the
scaffold of a vascular graft is made of a slowly
disappearing, biocompatible mesh. The challenge is to
design tubular fabrics which at the time of
implantation display the mechanical strength,
suturability and limited thrombogenicity of the
standard arterial grafts, but lead, after the
disappearance of the synthetic polymers, to the
formation of a regenerated tissue structure of
appropriate strength and hemocompatibility, without at
any time allowing the rupture of the arterial wall.
Standard dimension aortic and carotid grafts made
entirely of bioresorbable polymers function
effectively in canine, bovine and porcine models, to a
point where the mechanical and biological properties
of the conduit are due entirely to newly formed
tissues, which can adapt to the growth of the
recipient.

Several investigators have formulated the research
hypothesis that the elements of a normal arterial wall may
be reorganized through natural repair processes if the
scaffold of a vascular graft is made of a slowly
disappearing, non-toxic mesh (1,2,3,4). The challenge is to
design tubular fabrics which at the time of surgical
implantation display the mechanical strength, ease of
handling, suturability, limited blood loss and low
thrombogenicity of the standard arterial grafts, yet lead,
after the disappearance of the synthetic polymers, to the
formation of a regenerated tissue structure of appropriate
strength and hemocompatibility. The development of the new
blood conduit involves the simultaneous decay of the
synthetic polymer and growth of the constitutive elements of
an arterial wall. The acid test for appropriate matching of
these two processes is the absence of aneurysm formation or

rupture of the prosthesis under the stress of transmural hydrostatic pressure (5).

A similar approach has been proposed for the replacement of other duct structures in the body (e.g. trachea, esophagus) or entubulation of regenerating tissue (e.g. nerve guidance channel), but much of the initial experience in controlling the production of new tissue in vivo has been derived from arterial replacement.

A distinction can be made between biodegradation and bioresorption as related to tissue engineering. Biodegradation is an in vivo depolymerization process where changes in implant shape, dimensions or mechanical integrity result from a breakdown which liberates macromolecular or particulate fragments. Whereas the original material may have been quite biocompatible at its interface with living tissue, debris and split products cannot be easily eliminated by the body, and are likely to be recognized as "non-self" by the host organism. In contrast, bioresorption implies that the living recipient of the implant is not only capable of depolymerizing the synthetic material, but will also eliminate it by phagocytosis, intermediate metabolism, consumption as a fuel substance, or excretion of the decomposition products. In other words, to demonstrate bioresorbability, the products of depolymerization must either be transformed along standard biochemical pathways, and then consumed locally without harmful effects to tissues, or they must be eliminated through standard excretion mechanisms, which typically involve coupling with water-soluble transport molecules.

Historically, the first step in the design of bioresorbable vascular grafts has been the search for polymer yarns which would depolymerize at the appropriate rate in the warm, watery environment of a mammalian body. The standard bioresorbable suture materials (polyglycolides and polylactides) which lose their mechanical strength within two or three weeks, often lead to catastrophic failure of the implants. Materials which last over six months often elicit an intense giant cell reaction. Yet there is no clear cut relationship between the stability of the polymer and the thickness of the tissue layer formed on the luminal side of a vascular graft, perhaps because the intensity of the tissue reaction is not only related to the chemical nature of the polymer, but also to the mass of material involved. Encouraging results have been obtained with polymers which degrade over a period of two to more than four months. Such materials can be novel bioresorbable homopolymers (e.g. polydioxanone) or copolymers, composites of polyurethanes (6) with polylactides (4) or fibrin (7), or fast resorbing polyglycolide yarns coated with "retardant" polymers (5), e.g. a mixture of polylactide and poly (2,3 butylene succinate).

In our own experience an 8mm internal diameter, 8-9 cm long aortic graft made entirely of bioresorbable materials has functioned effectively as an arterial conduit in dogs for as long as six months. Thus far the best results, measured in terms of patency, absence of aneurysmal dilation, and formation of a new blood interface devoid of mural thrombosis and intimal hyperplasia, have been observed with polymers which have largely decayed at three months, as ascertained by standard and polarized light microscopy, as well as determinations of molecular weight by gel permeation chromatography (4,8). Since at that point the synthetic fabric has lost its structural strength completely, the newly-formed tissues provide the mechanical integrity of the conduits. However the healing process has not yet stabilized and a true endpoint has not been reached from a cell biology standpoint. Interestingly enough, we have never observed the incidence of infection with fully bioresorbable prostheses. Stress-strain studies performed on the tissue formed on the inside of the original implant indicate that this element of the conduit, which morphologically appears to be made of fibroblasts, myofibroblasts and collagen fibrils, has about one quarter to one half the mechanical strength of a control specimen from the same animal (3). The experience of other investigators suggests that the compliance of the original prosthesis is a critical factor in obtaining the development of a circularly-oriented smooth muscle layer and the deposition of elastin (4). Aortic implants placed in young growing animals have shown the ability of the newly-formed conduit to double in size to match the growing diameter of the native artery (9). It seems also that the use of bioresorbable suture material in matching the anastomoses is a critical factor in preventing the formation of stenosis at the point of attachment of a biodegradable prosthesis with the aorta in a fast-growing animal such as the miniature pig. We have observed calcium salt deposition in the interstices of a non-woven composite fabric made of polyurethane and fibrin, implanted as an aortic graft in the rat, but neither our group, nor others have yet reported calcification with prostheses made of a woven or knitted bioresorbable fabric.

TISSUE ENGINEERING

Now that the research hypothesis of controlled in vivo synthesis of living tissue has received its initial validation, what can be the contribution of "tissue

engineering?" I will define tissue engineering as the
creative concept of locating at the level of tissues -
meaning assemblages of different cell types and their ground
substances or secretion products - the interface between
engineering/materials science with modern biology.
    The issue, as I see it, is how to reconcile the
engineering analysis and systems approach to life processes,
both of which adequately describe the phenomena, but have
yet no obvious link with the underlying cellular and
molecular processes, with the pathobiology approach to organ
dysfunction, which is increasingly served by a cellular and
molecular approach, and focuses on the resolution of
clinically relevant problems, but features a number of
unquantifiable "black boxes" which defy engineering
analysis.
    If tissue engineering is to be the meeting ground of
engineering and materials science with cell biology, both
disciplines need to develop new methodologies to effect a
fruitful encounter. For engineering, the challenge is first
one of instrumentation to measure the physical properties of
tissues on very small samples and under conditions which do
not lead to post-mortem alterations. Soft tissue
micromechanics (3) must address the isotropy and uniformity
(or absence thereof) of the components of regenerated
tissues, as compared to normal tissues. The next challenge
is to develop computer models of these structures which may
help to predict failure mode and the role of individual
building blocks. For materials science, the question is to
relate surface and bulk characteristics, including the
microarchitectonics of a bioresorbable tubular fabric, with
the intensity and duration of the cellular response to the
implanted material. For cell biology, the challenge is two-
fold: to address the various components of the tissue
reaction in quantitative terms, using the methods of
computer-assisted histological and histochemical analysis;
and to find ways to influence actively the individual steps
of tissue regeneration (10), using either cell seeding with
endothelial cells (11) or smooth muscle cells (12), drug
release systems, immobilized enzymes, growth activators or
growth inhibitors attached to the bioresorbable polymer
scaffold. Throughout this effort, it will be critical to
express design objectives in biologically meaningful terms.
    For vascular wall regeneration, the issue of animal
model deserves serious consideration. Long term experiments
in large adult animals are prohibitively expensive.
Therefore the number of prostheses in any cohort is of
necessity small. Explant study methods which consume all or
large parts of a specimen cannot be statistically correlated
with data from morphology, histology or biochemistry except
in large scale studies. Species differences also leave

unresolved questions as experimentalists diversely use rabbits, dogs, sheep, baboons and miniature pigs. Our group has recently elected the microsurgical implantation of aortic prostheses in the rat as a screening method for new materials or new doping substances on existing materials.

We subscribe to the view that the formation of a new arterial wall in relation to a bioresorbable polymer scaffold represents a special case of the wound healing process. Contrary to our initial hypotheses, the core of the wound healing process does not take place within the interstices of the porous, bioresorbable fabric, but inside and outside of it, with the result that the disappearing polymer strands are progressively squeezed between a neointima and neo-media on the luminal side, and a neo-adventitia on the outer side. The various phases in the cellular reactions to a bioresorbable implant are qualitatively the same as with a biodurable material. To control the process and guide it toward a structure as closely resembling the natural arterial wall as possible, it appears important to dampen the intensity of the inflammatory reaction, to avoid the retention of a permanent irritant, to minimize the proliferation of fibroblasts and the deposition of collagen fibrils, to favor the differentiation of myocytes and their spatial organization in circular and longitudinal bands, and lastly to achieve a continuous endothelial lining on the luminal side. Competing with nature is a tall order, but given time and ingenuity, the goal of a biological replacement for a diseased artery no longer appears totally unrealistic.

REFERENCES

1. Bowald S, Busch C, Erikson I (1980). Absorbable material in vascular prostheses: A new device. Acta Chir Scand 146:391.
2. Greisler HP, Kim DV, Price JB, Voorhees AB (1985). Arterial regeneration activity after prosthetic implantation. Arch Surg 120:315.
3. Richardson PD, Parhizgar A, Sasken HF, Chin TH, Aebischer P, Trudell LA, Galletti PM (1986). Tissue characterization by micromechanical testing of growths around bioresorbable implants. In Nosé Y, Kjellstrand C, Ivanovich P (eds): "Progress in Artificial Organs 1985." Cleveland: ISAO Press.
4. van der Lei B, Wildevuur CRH, Nieuwenhuis P (1986). Compliance and biodegradation of vascular grafts stimulate the regeneration of elastic laminae in neo-

arterial tissue: An experimental study in rats. Surgery 99:45.

5. Galletti PM, Aebischer P, Sasken HF, Goddard MB, Chiu TH (1988). Experience with fully bioresorbable aortic grafts in the dog. Surgery 103:231.
6. Gogolewski S, Galletti G(1986). Degradable, microporous vascular prosthesis from segmented polyurethane. Colloid & Polymer Sci 264:854.
7. Soldani G, Palla M, Giusti P, Parenti G, Marconi F, Marchetti G, Bianchi G, Simi V (1987). A new fibrin-containing small diameter arterial prosthesis. Life Support Systems 5(1):9.
8. Aebischer P, Sasken H, Chiu TH, Goddard M, Galletti PM (1986). Totally bioresorbable grafts: Tissue reaction versus chemical composition. Life Support Systems 4(2):112.
9. Gogolewski S, Galletti G, Ussia G (1987). Polyurethane vascular prostheses in pigs. Colloid & Polymer Sci 265:774.
10. Noishiki Y (1978). Patterns of arrangement of smooth muscle cells in neointima of synthetic vascular prostheses. J Thorac Cardiovasc Surg 75:894.
11. Herring M, Baughman S, Glover J, Kesler K, Dilley R, Gardner A (1984). Endothelial seeding of Dacron and polytetrafluoroethylene grafts: The cellular events of healing. Surgery 96:745.
12. Yue X, van der Lei B, Schakenraad TM, van Oene GH, Kuit JH, Feijen T, Wildevuur CRH (1988). Smooth muscle seeding in biodegradable grafts in rats: A new method to enhance the process of arterial wall regeneration. Surgery 103:206.

Tissue Engineering, pages 25–30
© 1988 Alan R. Liss, Inc.

DEVELOPMENT OF A SMALL DIAMETER, COMPLIANT VASCULAR PROSTHESIS

Philip Litwak, Robert S. Ward, A. Jean Robinson, Iskender Yilgor, and Carol A. Spatz

Thoratec Laboratories Corporation
Berkeley, CA 94710

ABSTRACT: Compliant vascular grafts have been fabricated from a unique polyurethaneurea. Bulk and surface properties dictated by the final product design were decoupled by independently developing two separate materials (one optimized for its bulk properties, the other optimized for biocompatibility). The final polymer consists of a blend of surface active copolymer and bulk material. Grafts are extruded as multilayered concentric tubes. Both in vitro and in vivo evaluations have been performed. In vitro testing includes compliance, burst strength, kink radius, ultimate elongation, suture pullout, and accelerated flex testing.
In vivo evaluation is accomplished by implanting 8 cm lengths in the carotid arteries of mature goats. Technetium radionuclide angiography is used to determine patency. After 113 days, explanted grafts have demonstrated excellent blood compatibility and no perigraft reaction. Maximum thickness of endothelial and subintimal tissues bridging anastomoses has been 15-20 microns and in some cases consisted only of a continuous monolayer of endothelial cells. Islets of endothelial cells have also been identified along the graft wall.

INTRODUCTION

Arterial prostheses have been commercially available for approximately 25 years. Due to either the inherent nature of their base material or to their fabrication process, currently available devices are virtually non-compliant. When these grafts, with

Supported in part by grant #2 R44 HL 34933-02 from the National Institutes of Health.

acceptable long-term patency rates in high flow situations, have been reduced in diameter and used for peripheral, low flow applications their patency falls dramatically.

Stiff vascular prostheses in the elastic arterial system have the potential to interfere with the transfer of pulsatile energy. This can cause flow separation, excessive shear stresses, and turbulence. Thrombosis, anastomotic disruption, and intimal hyperplasia have been associated with mismatched compliance (1).

METHODS

Biomaterials

The basic biomaterial for this application is a segmented polyurethaneurea designated BPS-215. This material has undergone extensive physicochemical characterization (i.e. tensile strength, 7500 psi; ultimate elongation, 970 percent; initial modulus, 1500 psi; Youngs modulus, 1777 psi; hydrolytic stability, after five autoclave cycles no significant change in modulus or tensile strength). Complete toxicologic evaluation has been performed. All tests including: agar overlay, Class VI USP, hemolysis, unscheduled DNA synthesis, Ames Salmonella mutagenicity, and mouse lymphoma cell mutagenesis assay have been negative.

The surface active component of the final material blend (designated SMA-300) is a segmented polydimethylsiloxane-based polyurethane. Synthesis occurs in in a process similar to that of the base material, but this polymer is recovered from solution and stored as a solid material. It is redissolved in the BPS-215 bulk solution prior to final filtration at approximately 1.0 percent by weight. The actual SMA-300 concentration may be varied depending on the surface-to-volume ratio of the configured article. Critical surface tension of the base material has been lowered from 26 to 22 dynes/cm with the addition of SMA.

Graft Fabrication

Grafts are extruded as 4 mm diameter multilayered tubes. The inner, blood contacting, layer presents an open foam face and is approximately 8 mils thick. A solid impervious layer, adherent to the inner foam, provides the graft with strength and prevents fluid transfer. It is approximately 3 mils thick. Prostheses have been made both with and without an outer textured foam layer that serves as a tissue anchor. When present it is approximately 8 mils thick.

In Vivo Test Protocol

In vivo evaluation is performed in 35-60 kg neutered male goats using 8 cm lengths as interpositional grafts. Animals are heparinized intraoperatively using a dose (2 mg/kg) calculated to produce an ACT of 600+ seconds (normal = 142 + 28 sec.). Heparin neutralization is not used. Anastomoses are performed end-to-end using a 6-0 polypropylene continuous suture and 2.5x magnification. Seven to 10 days postsurgery 99m technetium radionuclide angiography is performed to assess patency. Periodic hematologic evaluations (CBC, platelet count, fibrinogen, platelet aggregation, and biochemical organ function tests) are performed during the observation period.

RESULTS

An eight year iterative process has led to our present reference graft. During this period over 50 different configurations have been evaluated. Fourteen grafts of this design have been implanted. All were patent at the time of the first angiogram (22-24 days postsurgery). . Six grafts were explanted after 23-33 days in vivo. The remaining eight grafts were harvested between 69 and 113 days after implantation. Three of these were not patent.

Explanted grafts were preserved in glutaraldehyde-formalin. All patent grafts showed the same response. None had the slightest visible thrombus at the anastomoses, nor was there evidence of gross thrombus adhered to the graft wall. A thin layer of white, glistening tissue covered, but did not obscure, each suture line. This tissue had grown onto the graft for: 4-5 mm after 23-33 days, 5-6 mm after 69 days, and 7-9 mm after 113 days. None of this smooth overgrowth was visible on the midgraft wall (Figure 1).

FIGURE 1. Explanted 4 mm diameter vascular prosthesis after 113 days observation (distal anastomosis to the left). Smooth, glistening, white tissue covers each anastomosis and is attached to the graft wall.

Each of the nonpatent grafts was surrounded by fibrotic response that had severely kinked them as it became organized. Within the lumen a solid thrombus completely filled and was attached to the textured inner surface.

Samples of explanted grafts were submitted for SEM and routine microscopic sectioning. Each test demonstrated a continuous layer of spindle–shaped cells growing from the cut arterial end onto the graft (Figure 2). In some cases there were no cellular layers beneath this apparent endothelial cell layer. In some cases a thin subintimal layer was attached to and grown into the textured surface with endothelium lining this tissue (Figure 3). Organized thrombus was not seen in areas beyond the immediate anastomoses. A few blood cells and fibrin strands were the only structures adhered to the wall, except for isolated islets of the spindle–shaped cells. These patches were usually 50–100 cells in size with little or no tissue between them and the graft surface.

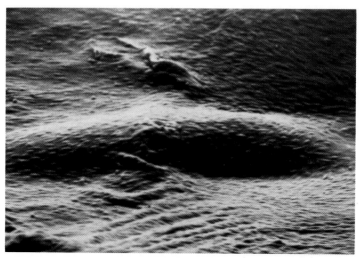

FIGURE 2. SEM of a distal anastomosis after 32 days implantation. A 6-0 polypropylene suture is visible beneath the continuous endothelial layer that has grown across the anastomoses.

In no case have we seen an inflammatory response at the subcutaneous tissue–graft interface. Sinus tracts have not been observed. Three of the thrombosed grafts were severely kinked by a fibrotic response. Further evaluation revealed that these were fabricated from a lower modulus material with increased moisture vapor transmission. This polymer allowed fluid and protein to pass through the graft wall, coagulate, and eventually fibrose.

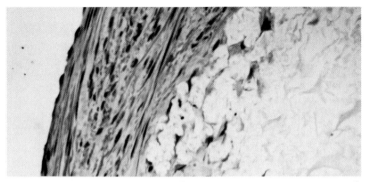

FIGURE 3. Photomicrograph of a proximal anastomosis after 32 days implantation. A thin (approximately 15-20 microns) layer of subintimal tissue and endothelium has proliferated across the anastomosis and is anchored in the graft's porous surface.

DISCUSSION

When developing polymers for biomedical use, it is seldom possible to optimize both the surface and bulk properties simultaneously in a single material. This frequently limits the developer's ability to tailor a product to a specific application. In homogeneous polymers each of these attributes is dependent on the same molecular variables, i.e. structure and weight.

Most methods of separating surface and bulk properties depend on expensive and nonpermanent surface treatments or coatings applied to the configured article. This polymer is unique because it consists of a blend of bulk material and surface modifying additive (SMA).   These SMA concentrate at the surface to significantly change the outer few molecular layers.   Since this is the region which determines biocompatibility, SMA can dramatically improve this property. The small amount of additive needed has virtually no effect on the bulk properties of the base polymer (2).

Matching the compliance of a prosthesis to natural artery is extremely difficult. Not ony are these values hard to measure, but there are disease processes that will alter compliance in localized areas (3).   The vast majority of studies relating patency and compliance have not addressed additional graft characteristics that could affect patency.   Surface chemistry, surface texture, wall thickness, and suture hole size will be different for each graft and can influence how long a prosthesis remains functional.

Lyman et al (4) and Seifert et al (5) reported on experiments designed to relate patency and compliance. They compared two types of polyurethane grafts, compliant and non-compliant, and found six of 11 compliant grafts (3-8 mm diameter) were patent at removal, whereas all six non-compliant grafts were thrombosed.

Compliance of a vascular prosthesis, no matter how closely matched to the arterial compliance, is not the sole determinant of patency. The blood contacting material must be biocompatible, the degree of which may vary depending on the type of flow surface. Traditionally, vascular grafts have had porous walls and textured surfaces. This allows a pseudointima to firmly attach to the surface and be nourished by ingrowing tissue and its accompanying vascular network. It is generally believed that this pseudointima is required if endothelialization is to occur (6). However, in man the significance of this observation is unknown since the occurence of endothelialization in vascular grafts has been questioned (7).

In summary, the preponderance of scientific evidence concludes that small diameter vascular grafts must be compliant for long-term patency. They must provide some type of textured surface for adhesion by the regenerating endothelial and subintimal tissues. We have described the configuration and initial testing of a small diameter, polyurethaneurea vascular graft that addresses each of the requirements of the ideal prosthesis.

REFERENCES

1.   Abbott WM, Cambria RP (1982). Control of physical characteristics (elasticity and compliance) of vascular grafts. In Stanley JC (ed): "Biologic and Synthetic Vascular Prosthesis," New York: Grune and Stratton.
2.   Ward, RS, White KA, Hu CB (1984). Use of surface-modifying additives in the development of a new biomedical polyurethane-urea. Colloquium for Polyurethanes in Medical Engineering, Denkendorf, West Germany.
3.   Hasson JE, Megerman J, Abbott WM (1985). Increased compliance near vascular anastomoses. J Vasc Surg 2:419.
4.   Lyman DJ, Fazzio FJ, Voorhees H, Robinson G, Albo D Jr (1978). Compliance as a factor effecting the patency of a copolyurethane vascular graft. J Biomed Mat Res 12:337.
5.   Seifert KR, Albo D Jr, Knowlton H, Lyman DJ (1979). Effect of elasticity of prosthetic wall on patency of small-diameter arterial prosthesis. Surg Forum 30:206.
6.   Clagett CP (1982). In vivo evaluation of platetlet reactivity with vascular prostheses. In Stanley JC (ed): "Biologic and Synthetic Vascular Prostheses,"New York: Grune and Stratton.
7.   Sauvage LR, Berger KE, Wood SJ (1974). Interspecies healing of porous arterial prosthesis: observations 1960-1974. Arch Surg 109:698.

Tissue Engineering, pages 31–35
© 1988 Alan R. Liss, Inc.

THE FUTURE OF VASCULAR PROSTHETICS

Vincent T. Turitto

Department of Medicine, Mount Sinai Medical Center
New York, New York 10029

The long-term goal in the field of vascular prosthetics
is simple: all grafts placed in the circulatory system will
be developed from readily available materials and these
prostheses will behave in a manner which is clinically
indistinguishable from the original healthy vessels they
replace for the life of the recipient. The steps necessary
to arrive at such a goal are not evident, however. In fact,
where the program will be in ten years is certainly not
clear. As a Co-chairman of recently held symposium
sponsored by the New York Academy of Sciences on "Blood in
Contact with Natural and Artificial Surfaces" (1), and a
participant in the initial symposium held approximately ten
years earlier on a similar topic (2), I have the opportunity
to assess progress over that ten year span. In the next few
pages I would like to summarize the advances during that
period in order to offer predictions for the next ten to
twenty years.

BIOMATERIALS

With respect to improvements in prosthetic materials
themselves, there has been no major breakthroughs and the
recent conference offered only token consideration to the
possibility that new materials might be substantially better
than previous available surfaces. The major problem remains
that small vessel diameter (< 4 mm) grafts fail with rapid-
ity when implanted in the vasculature and such prostheses
are used only in the absence of available homologous vessels
(3). The reasons underlying the failure of small vessel

This work supported in part by Grants HL 38933 and HL
35103.

replacements are related to thrombosis and cellular proliferation (intimal hyperplasia). It is in these two fields that progress in our understanding of basic events has been fastpaced compared with the development of novel biomaterials.

THROMBOSIS

The field of thrombosis has been particularly active. Mechanisms of platelet attachment to and aggregate growth on damaged vascular surfaces have been clarified both with respect to the biochemical role of certain glycoproteins on the platelet and vessel wall and the influence of blood flow in modulating such deposits (4). More recently, the role of platelets in coagulation events on such surfaces is being approached (5). The specific mechanisms by which coagulative events are influenced by cellular interactions in flowing blood will certainly be an area of development in the next decade. However, much of the information which has been derived for vascular surfaces will need to be applied to prosthetic materials where entirely novel mechanisms may be initiated. The influence of specific prosthetic surfaces on fully active platelet and coagulation systems has not been investigated to any great extent, especially when many of the interactions on the prosthetic surface occur in the vicinity of an injured vascular surface. Specific vascular promoters of interest in the coming years will be tissue factor (coagulation), collagen (platelet activation) and such glycoproteins as von Willebrand factor and fibronectin. The extent to which feedback mechanisms, operating between the artificial and vascular surface, may engender synergistic activation of blood elements is certainly an area of interest. Systems of controlled complexity will need to be developed to explore the interactions among prosthetic surfaces, injured vessel wall, blood coagulation, platelet activation and fibrinolytic mechanisms in order to appropriately assess the relationship between surfaces and thrombosis.

The application of such studies to clinically relevant thrombosis is considerably more difficult. Basic research studies are essentially short term (of the order of seconds to days), whereas thrombosis on vascular prostheses is long term (months to years). No models for investigating such long term events are currently available which are not expensive and essentially single time probes (implantation with termination of the experiment at specified time points). Few ideas concerning the potential differences between acute and chronic thrombosis have evolved and the models to test such differences are not available. Continuous monitoring of thrombosis in experimental models is an obvious means of obtaining such information and non-invasive methodology is being currently developed (6) but is not likely to be sensitive enough in its evaluation until the next century.

CELL PROLIFERATION

The past ten years have seen tremendous advances in the field of cell biology. Culturing of human endothelial cells, the natural lining of blood vessels, was achieved only a little more than a decade ago (7), and human aortic endothelial cells only in recent years. Such progress has provided the impetus for creating cell-lined prosthetics which would be passive to the elements of blood (See Jarrell, this volume). While such an approach is attractive, a major finding of the recent decade must also be considered: endothelial cells do not provide solely a passive barrier; they are active cells which synthetize a host of pro-and anti-thrombotic elements. Thus, it is likely that simulation of the natural vessel will not simply involve coating the cells on a surface. Endothelial cells will have to be in an appropriate environment whereby they will remain both viable and functional. The normal functioning of endothelial cells involves communication with different cell types through various regulating signals which may arise from the blood itself or the deeper layers of the vessel wall. Unlocking the secrets of vascular cell communication is of broad interest to a variety of biological functions and will be a project that will extend well into the 21st century. In the case of vascular implants, progress may be aided by the similarity between the intimal hyperplasia observed in grafts and the smooth cell proliferation which is typical of the early stages of atherogenesis (8). A past failure of investigations in the area of vascular prostheses has been that these studies often have been conducted in isolation from many of the biological sciences. Thrombosis and cell proliferation are problems which extend into a wide variety of disciplines of both a clinical and basic nature. The parallels with the much more pervasive disorder, atherosclerosis, should bring new insight into the mechanisms of failure, and consequently the means of prevention.

ROLE OF ENGINEERING

Engineers have been and will continue to contribute to our understanding of blood-surface interactions; their role is likely to grow as the influence of physical forces on biological function becomes more apparent. This brief summary has emphasized the biological aspects associated with the failure of prosthetic vascular replacement. It is in this area that growth in our understanding of cellular mechanisms has been dramatic over the past decade, whereas the role of fluid dynamics and mass transport, while not so rapid, has been steady. Presently, a number of areas of traditional interest primarily to engineers, are now incorporated into well-designed biological studies. The realization by biologists that many cardiovascular events occur only in the presence of blood flow and that such flow actually modulates the biological outcome is likely to lead to closer collaborations of engineers with traditional biological scientists in the coming decade. Such interdisciplinary teams will be characteristic of the best research groups of the 21st century.

At present fluid mechanics and mass transport are closely involved in such areas as:

1. platelet interactions with artificial and biological surfaces,

2. the growth of platelet thrombi in recirculated flow regions,

3. cell entry in branched microvessels,

4. the influence of aortic stenosis on thrombosis,

5. the modulation of fibrin deposition by local shear conditions,

6. morphological and functional effects of shear stress on endothelial cells,

7. the growth/embolization of platelet aggregates in flowing blood,

8. the influence of shear stress on platelet/white cell interactions.

This list obviously reflects the bias of the author; however, in general, platelets and by endothelial cells have been the most studied cells with respect to blood flow. What has been less studied and will undoubtedly be the subject of future research is the influence of flow on the coagulation enzymes; the extent to which pressure/flow relationships can modulate cell proliferation, and the overall regulatory effect of flow on cellular synthesis and enzymatic function. These are all topics of the decades ahead.

REFERENCES

1. Leonard EF, Turitto VT, Vroman L (eds.) (1987). Blood in Contact with Natural and Artificial Surfaces, Ann. N. Y. Acad. Sci., Vol. 516.

2. Vroman L, Leonard EF (eds) (1977). The Behavior of Blood and its Components at Interfaces, Ann. N. Y. Acad. Sci., Vol. 283.

3. Callow AD (1982). Historical overview of experimental and clinical development of vascular grafts. In Stanley TC (ed): "Biologic and Synthetic Vascular Prostheses," New York: Grune and Stratton, p 11.

4. Turitto VT, Baumgartner HR (1987). Platelet-surface interactions. In Colman RW, Hirsh J, Marder VJ, Salzman EW (eds). "Hemostasis and Thrombosis: Basic Principles and Clinical Practice," Philadelphia: J. B. Lippincott Co., p 555.

5. Weiss HJ, Turitto VT, Baumgartner HR (1986). Role of shear rate and platelets in promoting fibrin formation on rabbit subendothelium. J. Clin. Invest. 78:1072.

6. Anon. Devices and Technology Branch Contractors Meeting 1987. Division of Heart and Vascular Diseases, National Heart, Lung and Blood Institute, Dec. 7-9.

7. Jaffe EA, Nachman RL, Becker CG, Minick CR (1973). Culture of human endothelial cells derived from umbilical veins: Identification by morphologic and immunologic criteria. J. Clin. Invest. 52:2745.

8. Lee KT (ed) (1985). "Atherosclerosis," Annual N. Y. Acad. Sci., Vol. 454.

Tissue Engineering, pages 37–38
© 1988 Alan R. Liss, Inc.

Summary/Discussion

Vascular Prosthesis

Vincent Turitto

In a brief overview of the field, Vincent Turitto described how the area of vascular prosthesis had long been identified with biomaterial development. However, no major breakthroughs in this area have occurred in the past ten years — certainly none that have reached clinical application. Novel approaches consisting of closer simulation of the synthetic and regulatory properties of the natural vessel wall currently appear to be accepted for the development of vascular replacements. In this light, work presented by Bruce E. Jarrell and Robert M. Nerem relating to endothelial cells is of particular interest. These investigators discussed their attempts to deposit cells on appropriate substrates and to investigate their responses to various flow perturbations

Pierre M. Galletti presented his work on dissolvable vascular prostheses. He has taken advantage of the tendency for ingrowth of various cells into vascular implants in an attempt to create a regenerated tissue structure of appropriate strength and hemocompatability. He reported success with several different polymer materials for up to six months in the aorta of a dog. The grafts remained open, and showed no indication of aneurysmal dilation, mural thrombosis or intimal hyperplasia. However, he stressed that the test period was not sufficient at present to determine whether the ingrowth processes lead to a stable vascular structure.

Reservations about the use of such techniques were expressed by Bruce Jarrell, who indicated that surgeons would be very concerned with the clinical use of such tissue structures because of the possibility of rupture in aortic locations. Vincent Turitto raised questions regarding combining techniques of seeding vascular cells with this novel implantation technique in order to obtain a more directed composition for ingrowth. Pierre Galletti indicated that this would certainly be possible, but would require considerable study.

Philip Litwak described a unique polymer (polyurethaneurea) graft in which a composite approach using two different materials was adopted to obtain optimal bulk and biocompatibility properties for vascular implants. Tubes implanted in the carotid artery of goats for as long as four months had excellent blood compatibility with no cellular deposits or thrombus formation on the graft, and some ingrowth of endothelial cells from the host artery. Diameters as small as 3 mm were utilized. Both a smooth and a rough version of the material were employed. Neither gave overt thrombus formation, but the textured version gave better adhesion of the endothelial lining to the host/graft interface and better longevity. Fabrication of grafts of different compliances showed comparable results.

Vincent Turitto emphasized that although patency and a clear graft surface were highly desirable, they were minimal criteria, since embolization of loosely adhering materials could be deleterious to distal organs. However, techniques for continuously monitoring embolization are not available at present.

The subject of animal models was a topic of vigorous discussion. There was general agreement that no animal model used at present was reasonably priced, easily handled, and similar to man in its behavior in all respects. Dogs were likely to be gradually phased out due to pressures from animal rights groups, but appropriate models may still exist with goats, pigs, and sheep.

Tissue Engineering, pages 39–44
© 1988 Alan R. Liss, Inc.

TISSUE MECHANICS OF BLOOD VESSEL WALLS

Peter D. Richardson

Division of Engineering, Brown University
Providence, RI 02912

ABSTRACT   There is increasing evidence that acute events which occur in blood vessels and lead to death or major disability are associated with mechanical failure of part or all of the thickness of the blood vessel wall.   Pathological studies in series having significant numbers of patients, and mechanical measurements of tissue properties combined with stress analysis of complete vessels, focus attention now on the existence of cracks and tears in vessel walls in locations which would be expected to have the highest stresses, with these stresses in the range where fracture of tissue is observed in *in vitro* experiments.   While various techniques for repair and replacement of vessel segments have been developed in the past few years, the subsequent rate of failure assures us that we do not yet have a sufficient understanding or control of the processes involved.   Improvements in tissue engineering in this area will require a combination of better understanding of the relationship between structure and mechanical properties of the soft tissues that constitute vessel walls, and approaches to repair and substitution that allow for the various changes and reactions that occur in tissue.

INTRODUCTION

There is an increasing body of evidence that the acute clinical events associated with blood vessel failures, which include heart attacks, strokes, aneurysms of cerebral arteries and of vessels in other locations, and dissecting aneurysms, begin very frequently with a local mechanical failure of a portion of the tissues that constitute the vessel wall. Studies by Davies and Thomas (1), Davies and Thomas (2), and Falk (3) have examined large cohorts of hearts taken from persons who experienced sudded death from myocardial ischemia. It was found that intimal fracture in coronary arteries was present in a considerable majority of the cases (over 85%). In these coronary arteries the fracture site was connected with a major thrombus, which was either totally occlusive or appeared to have embolized sufficient thrombotic material to account for the fatal myocardial ischemia, with invasion of some of the thrombotic material into the cracked portion of the tissue. These series of studies have given statistical weight to what had been previously only anecdotal evidence.

For several years there has been an increasing level of concern about the high incidence of failure of vascular grafts, be they autologous transplants or prosthetic segments of materials which have been developed as early response in tissue engineering. Some persons have suggested that the typical mismatch between the mechanical behavior of the retained native vessel segment and the graft predisposes the combination to failure. Examination of aortas at post-mortem show the tendency for an increased level of local breakdown of tissue continuity in the lumen of the vessel as the age of the subject becomes greater.

Instrumentation currently available does not allow us to make highly localized measurements of stress and deformation within blood vessels under *in vivo* conditions. We have approached the problem of stress analysis in blood vessels, and the adequacy of tissue replacements, including bioresorable materials used for the construction of vascular grafts, by developing techniques to perform micro-mechanical tests on small component pieces of blood vessels; taking measurements of typical properties and incorporating them in finite-element-based methods for analysis of stress and deformation; and comparison of the mechanical behavior of segments of tissue with the histologically-observable structural components and biochemical constitution.

MEASUREMENTS OF TISSUE MECHANICAL BEHAVIOR

In our laboratories we have investigated the mechanical properties of the three layers (intima, media, and adventia) of various vessels, the samples taken in the axial and circumferential directions, and occasionally at 45°. Our studies include some for carotid arteries, Owens and Richardson (4), tissues developed around bioresorbable implant materials, Richardson et al. (5), Richardson (6), for coronary arteries, Keeny and Richardson (7), and for the aorta, Born, Davies, Lendon, and Richardson (8). These measurements show that there is considerable variation in mechanical properties of vessels between individuals of similar age, similar to the non-invasive ultrasound studies of the aorta, Kalath, Tsipouris, and Silva (9). These ultrasound studies, of course, measurement integrated effect throughout the vessel. There is also considerable variation between the mechanical properties of the different layers of the vessels in an individual at different locations on those vessels, sometimes even at the same axial position along the vessel. A common feature in the mechanical response of these tissues is a high compliance at small deformations, but as the deformation increases, the incremental load required for further deformation increases more rapidly. Where plaques have been studied after removal from vessels, the cap tissue is generally less compliant than that of the nearby normal intimal tissue; in some cases it was up to about 5 times stiffer. Intima also exhibits considerable anisotropy, with the tissue stiffer in the axial direction than in the circumferential direction. The basic mechanical tester with which we began our experiments is illustrated in the paper by Richardson et al. (5). We have a similar instrument operating in London, but with a computerized real-time data acquisition system for observation of the elongation of the tissue samples.

The fracture process in tissues taken from blood vessels appears to be quite complicated. The use of several fiduciary marks placed on tissue in order to observe the local deformations has shown that local failures by some sort of snapping process can be seen in tissues before they fail completely, with full separation of two parts of a test specimen. These little internal failures in tissue specimens can occur at quite low stresses, even below 50 kPa, and this is a stress level which is frequently achieved and exceeded under normal physiological conditions. These observations suggest that there may be some degree of highly-localized and small-scale internal fracture of arterial wall tissue occurring regularly, and one may suppose that there is also some repair process which occurs. Stresses required for completed fracture of intimal tissue have ranged from the order of 150 kPa up to 1000 kPa. From the practical point of view it appears that the ultimate strength of such tissues has a considerable range, which may account for the variable incidence of vessel fracture amongst individuals who have similar ranges of arterial pressure loading their vessels.

COMPUTATION OF VESSEL WALL STRESS DISTRIBUTIONS

Finite-element methods can be used to calculate wall stress distributions. There are certain overall conditions which need to be satisfied in the computations. One is that the radial integral of the circumferential stress across the wall, for a unit axial length of the vessel, should balance the blood pressure multiplied by the internal radius of the vessel. Another condition is that the local deformations of the tissue must be mutually compatible. The vessel wall is divided into layers radially in its natural state (although the structure can undergo considerable modification as a consequence of the developement of atherosclerotic lesions), and the outer deformation of one layer needs to match the inner deformation of the next layer as one moves radially outwards. In a typical situation the innermost layer has a larger circumferential elongation than the outermost layer, because the innermost layer reduces its thickness, and the elongation ratio of the next layer in the radial direction is less, because it does not have to extend as far in order to match the outer circumference of the inner layer. This relationship continues throughout the wall. When one finds segments of layers which are relatively stiff, as may happen due to additional cross-linking or to calcification, one can expect that the tissue will behave somewhat similarly to what one finds in fiber-reinforced materials. In such cases, loading of the stiff fiber occurs through shear deformation of the softer and more compliant material that surrounds it, especially near the ends of the fiber.

Computations have been performed for many specific cases of vessel walls. The first step is to select the geometry of the domains within a vessel wall to which different and specific mechanical relations will be applied. Each element within a finite-element structural model can be assigned its own specific stress-elongation relation. In many of our calculations we used either a symmetric half-artery model, or even an antisymmetric quarter-artery model, with up to twelve radial layers. The use of half or quarter artery models has been for the normal purpose of taking advantage of symmetry and reducing the total number of nodes for which computations have to be performed.

Computations for arteries with athersclerotic lesions occupying a finite arc within the vessel wall have led to some quite striking results. When the plaque material is relatively stiff (as with calcified plaque), the presence of the inclusion tends to reduce stresses in the neighboring intima. By contrast, a lipid pool tends to cause a significant increase in the maximum stress which is experienced in the intima. Depending somewhat on the precise details of the geometry and the arterial pressure applied, the maximum stress within the arterial wall is found either near the edge of the crescentic plaque, or in the middle of the intimal cap over the plaque, and comparison with an unpublished retrospective series examined by Professor Davies indicates that in the clinical cases the fratures occur most frequently at those sites. What is even more striking in the numerical results is that the increase in the maximum stress experienced in the intima is enhanced even more by an increase in stiffness of the plaque cap, within the range found typically in experiment, than is found simply by an enlargement of the arc of the vessel wall which is subtended by the region filled with lipid. This suggests that in some sense the increase in the stiffness of the plaque cap is a greater risk factor for mechanical failure and rupture of the plaque cap than, simply, the presence of some lipid within the vessel wall but under a cap having mechanical properties close to those of the nearby intima that is unaffected by the presence of the plaque.

Many different relations have been proposed for the representation of the nonlinear relationship between stress and deformation of soft tissues. We find that many of them can be fitted with closely similar accuracy to the experimental results of uniaxial loading. Comparison of possible relative merits of different formulations appears to be less significant at this point than investigation of the relationship between the observed mechanical properties and the structure of the tissue itself.

RELATION BETWEEN MECHANICAL BEHAVIOR AND TISSUE STRUCTURE

Collagen and elastin are the principal structural proteins of vessel walls. The different architectures of incorporation of collagen and elastin in the different layers of the vessel wall have been described with increasing precision over the past few years. However, no simple relationship between the structural protein contents of walls and their observed mechanical behavior has emerged. While the mechanical behavior of elastin has sometimes been compared to that of rubber, with the corresponding expectation of possibly simple analytical representation of mechanical behavior, it is clear that this is a somewhat fascile comparison and, moreoever, that collagen behaves very differently. Polymer science methodology has not been as illuminating yet as was initially hoped. Cross-linking of collagen has been identified as playing a major role in determining the mechanical strength of fibrous collagens. It has been supposed that several diseases of connective tissues are due to the defective organization of collagen, resulting in reduced tensile strength of the corresponding tissue. In the Marfan Syndome, collagen and elastin cross-linking defects have been suspected for many years, Abraham et al. (10). A reduction of the number of cross-links per collagen molecule reduces the tensile strength and increases the susceptibility of collagen fibers to degredation by collagenase (Vater, Harris and Siegel, 11). Pyridinoline is a collagen-specific covalent cross-link. It can be quantified in extracts of tissue using a non-equilibrium inhibition ELISA. This method was applied by Anne Whittle et al. (12) in investigation of lesions in human aorta that may predispose to dissecting aneurysms. However, there were no significant differences in the amounts or concentrations of pyridinoline in aortas with dissecting aneurysms compared with normal tissue. Investigations of other possible chemical relationships, including that to calcium, so far have not proved to give a very clear indication of the relationship between the mechanical properties and the observable constituents. However, it should be recognized that a number of the biochemical techniques require sample sizes at present which are fairly large compared with the size of the piece of tissue on which the mechanical measurements can be made. Thus, applications of these techniques at present may tend to disguise correlations that could be found if methods which require smaller amounts of tissue could be applied. One of the needs for advance in this area is improvement of assay techniques to allow small sample sizes to be treated.

SUMMARY

The increasing recognition of the importance of local mechanical behavior of soft tissues, coming from quantitative pathological studies, encourages greater efforts to be applied to research into mechanical behavior of these tissues, the relationship of variations in mechanical behavior to stress distributions, and investigation of the relationship between structure and organization of the tissue, and the observed mechanical response. Greater knowledge in this areas should allow far more successful methods to be developed for the repair and replacement of damaged soft tissues.

REFERENCES

(1) Davies MJ and Thomas T (1981). Phil. Trans. Roy. Soc. London, Series B. 294:225.

(2) Davies MJ and Thomas AC (1985).   Brit. Heart J. 53:363.

(3) Falk E (1983). Brit. Heart J. 50: 127.

(4) Owens C and Richardson PD (1985).   Proc. 11th Annual NE Conference on Bioengineering, pp. 12-15.

(5) Richardson PD, Parhizgar A, Sasken HF, Chiu TH, Aebischer P, Trudell LA, and Galletti PM (1986).   Prog. Artif. Organs-1985, Cleveland: ISAO Press, p. 1015.

(6) Richardson PD. Bull. NY Acad. Med. (to appear, 1988).

(7) Keeny SM and Richardson PD (1987).   Proc. 9th Annual Conf. Engineering in Engineering and Biology Soc., IEEE 3:1484.

(8) Born GVR, Davies MJ, Lendon C, and Richardson PD. Abst. for 10th Congress of the European Soc. Cardiology, Vienna, 1988.

(9) Kalath S, Tsipouis P, and Silva FH (1986).   Annals Biomed. Engrg. 14:513.

(10) Abraham PA, Perejda AJ, Carnes WH, and Uitto J (1982).   J. Clin. Invest. 70:1245.

(11) Vater CA, Harris ED, and Siegel RC (1979). Biochem. J. 181:639.

(12) Whittle MA, Robbins SP, Hasleton PS, and Anderson JC (1987). Cardiovascular Res. 21:161.

Tissue Engineering, pages 45–50
© **1988 Alan R. Liss, Inc.**

CELLULAR GROWTH IN SOFT TISSUES
AFFECTED BY THE STRESS LEVEL IN SERVICE

Y. C. Fung

Department of AMES/Bioengineering
University of California, San Diego
La Jolla, California 92093

ABSTRACT Natural tissues hypertrophy or resorb when
the working stress level exceeds beyond or falls
below a certain range. Heart muscle and blood vessel
wall hypertrophy when the blood pressure is raised.
Artificial tissue based on living cells would probably obey the same rules. The stress level acting on a
tissue is not only relevant to tissue strength and
integrity, but also to its stability in size, mass,
shape, and function through growth or resorption. A
quantitative measure of a manifestation of internal
structural changes of cellular and interstitial
growth is the residual strain at no-load condition.
It is one of the key parameter of tissue engineering.

INTRODUCTION

In cell culture, one sees rapid growth until a
confluent layer is formed. Then the culture changes very
slowly. Cells contiguous with neighboring cells are not
free to move - in fact, not even free to deform. Their
growth would be very subtle. An important question is:
how to detect growth in this situation, how to quantify
it, how to measure it?
The question is not peculiar to artificially engineered tissues, but to natural tissues as well. Growth
and change in living tissues is of general interest to
biology because it affects the stress and strain in
organs. Conversely, stress and strain has effect on
growth and change. Health of any organ, and thus of an
individual, will depend on a proper range of stress and

strain in homeostatic condition. Furthermore, pathogenesis is often associated with cellular or interstitial
growth, e.g., in atherosclerosis, cancer, hypertension,
hypertrophy. By extension, tissue engineering, the
controlled synthesis of a living tissue, must pay attention to the stress and strain levels in which the tissue
is grown and in which it is used. To detect and determine
the spatial distribution of tissue growth and change in a
whole organ is very difficult. This is why we believe that
our recent investigation of residual strains in living
organs is of interest (Fung, 1984; Chuong and Fung, 1986;
Liu and Fung, 1988). Residual strains are strains left in
an organ when all the external loads are removed. It can
be revealed by measuring the local deformation of the
tissue in an organ that occurs when the tissue is reduced
from the no-load condition of the organ to the zero-stress
condition of the tissue. This is done by cutting up the
organ in an appropriate manner. Residual strains exist in
most organs. If cellular or interstitial growth takes
place in an organ, the residual strains will change. By
measuring the change of residual strains and tracing the
cause of the change theoretically, one can determine the
quantitative aspects of the growth.

GENERAL CONCEPTS

The phenomenon of residual stress can be illustrated
by the following example: Isolate a segment of an artery.
Let it float in a neutrally buoyant physiological saline.
It is in a no-load condition. Take a pair of small
scissors and cut the artery along an axial plane. One
sees that the arterial wall springs open immediately as if
a tightened spring was released. The opened configuration
changes a little in the next few minutes as in viscoelas-
ticity. In about 15 minutes it reaches an asymptotic
steady-state "zero-stress" condition. If papaverine, a
smooth muscle relaxant, were added to the saline the
degree of opening and its time course remains unchanged
(see photographs in Fig. 2.9:4, p. 59, in Fung, 1984).
Hence the phenomenon just described is not due to smooth
muscle contraction or relaxation.

Similar observations have been made on the left ventricle, see figures on p. 60 in Fung (1984). More recent (unpublished) studies of the left ventricle have shown that thin longitudinal and latitudinal slices may need more than one cut to achieve a zero-stress state.

In engineering, residual stress has been taken advantage of in the making of gun barrels, prestressed concrete, etc., and in the analysis of thermal stress.

In biology, a cell is enclosed in a cell membrane; it grows by movement of its membrane; thus the vectorial displacement of a point on a surface depends on the orientation of the surface. Hence growth of a cellular continuum must be described by a tensor of rank 2. This consideration leads naturally to the concept of residual strain tensor, which serves to describe how cells are packed into a continuous tissue at no-load condition.

THE ZERO-STRESS OF NORMAL AORTA OF THE RAT

Sketch of a typical aorta is shown in Fig. 1. If a thin slice is cut from the aorta by planes perpendicular to the centerline of the vessel, the slice looks like a ring. If the ring is cut, it opens up into a sector as shown in Fig. 1. For geometric description we locate the mid-point of the arc of the inner wall of the vessel and draw two radii from that point to the outer tips of the ends of the sector. The angle between these two radii is defined as the sector angle and denoted by θ.

Rat aorta was sectioned successively into slices of approximately 1 mm thickness, cut as described above at one of the four points: I (inside), O (outside), A (anterior), P (posterior). Results are presented in Liu and Fung (1987). It is shown that for the normal rat aorta, θ is not a constant, but varies along the tree. The sector angle θ is about 160° in the ascending aorta, 60° in the carotid artery, zero in the mid thoracic aorta region, somewhat negative in a region including the diaphragm, then increases again to over 60° in the ilio lumbar region.

CHANGES FOLLOWING ACUTE INCREASE OF BLOOD PRESSURE

Hypertension was created in rats by banding the abdominal aorta with a metal clip placed right above the celiac trunk. The area of the restricted opening of the aorta was about 5% of the normal. At a planned time after the surgery, blood pressure of each rat was measured and the rat was sacrificed and the whole aortic tree was carefully separated, excised, and sliced. Then one cut was made which resulted in the opening of the section as shown in Figure 1.

The clipping of the artery was a powerful disturbance to the rat's circulation. As a result, the blood pressure in the upper body above the metal clip rose, at first rapidly, at a rate of about 10 mmHg per day, then more gradually, tending to an asymptotic exponentially. The mean pressure in thoracic aorta of normal rat was about 137 mmHg. In 40 days after the operation the mean thoracic pressure was 225 mmHg. Fifty percent of the total rise in blood pressure was achieved in about 7 days. The blood pressure below the clip (normal average 135 mmHg in ciliac artery) decreased at first, in 7 days it recovered to the normal value, then it rose gradually to an asymptotic value of 158 mmHg. In the meantime, hypertrophy proceeded, the blood vessel wall thickness increased, while its inner radius decreased, at first rapidly, then gradually. The rate and amount of thickness and radius changes varied with the location along the aortic tree. Fifty percent of the total thickening was obtained in 3 to 5 days. That the changes of blood pressure, blood vessel wall thickening, and vessel lumen radius were kept more or less in step is shown by the relative constancy of the mean hoop stress in the blood vessel: $\langle \sigma_\theta \rangle$ = pressure x inner radius $\div$ wall thickness, in the days following the operation.

The most immediate and profound change that occurred following aortic clamping kay in the residual stress as revealed by the change of opening angle when the vessel was cut. Figure 2 shows the change of opening angle in the ascending aorta following the operation. It is seen that the angle increased from $171°$ to $214°$ in 4 days, then decreased as rapidly back to $171°$, followed by a more gradual exponential decrease to an asymptotic value of $126°$. The total swing of the angle of opening was as large as $88°$ in the ascending arota. In other regions of

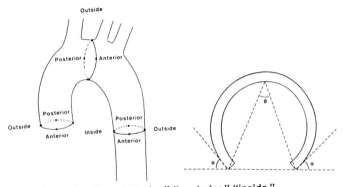

Fig. 1   Left: Nomenclature for sites: "anterior," "posterior," "inside," and "outside." Right: Definition of the opening section angle θ and the sum of the angles between the tangents to the vessel section at the site of the cut and the x-axis, α.

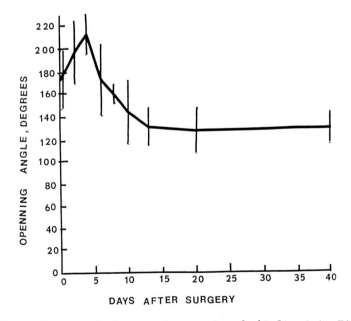

FIG. 2 —   Change of the opening angle, θ (defined in Fig. 1), when the artery is cut in a longitudinal plane, with the number of days following the celiac arterial clamping operation. Mean ± S.D. (n = 6).

the aorta the total swing was smaller and the course of change varied with the location on the aortic tree.

RELEVANCE TO SOFT TISSUE ENGINEERING

In controlled growth of cells into a living tissue, the stress and strain among the cells is one of the determinants of the tissue. Residual strain reveals the mechanical relationship among the cells. It is large and easy to measure in cases such as the blood vessel. It varies with the load that acts on the tissue. In engineering the tissue, one should simulate the stress to which the tissue is going to be subjected when it is implanted in vivo. Alternatively, one must be assured that when the tissue is removed from the culture environment into the human body, adaptation can be successively achieved. The later is, of course, a part of the engineering problem.

ACKNOWLEDGMENTS

This research is supported by NSF through grant No. EET 85-18559.

REFERENCES

1. Fung YC (1984). Biodynamica: Circulation. Springer-Verlag, New York.
2. Chuong CJ, Fung YC (1986). Residual stress in arteries. In Schmid-Schönbein GW, Woo, SL-Y, Zweifach BW (eds): Frontiers in Biomechanics, New York: Springer-Verlag, pp 117-129.
3. Liu SQ, Fung YC (1988). Zero-stress states of arteries. J Biomech Eng 110: 82-84.

Tissue Engineering, pages 51–56
© 1988 Alan R. Liss, Inc.

CELLULAR HYPERTROPHY CAN BE INDUCED BY CYCLICAL MECHANICAL STRETCH IN VITRO[1]

Louis Terracio*, Walter Peters**, Brian Durig**, Bonnie Miller***,Karen Borg*** and Thomas K. Borg***

*Department of Anatomy, **Department of Mechanical Engineering, ***Department of Pathology, University of South Carolina, Columbia, SC 29208

INTRODUCTION

Most cells in multicellular organisms are continually subjected to a variety of extracellular stimuli. Cells have developed a wide variety of mechanisms to perceive and respond to these stimuli. In addition to the well studied chemical signals, physical parameters can also serve as extracellular signals. The mechanisms for sensing these extracellular signals, their transduction and ultimately, the generation of a cellular response, have evolved into precisely regulated events by the cell. Investigations on how cells perceive and response to various stimuli have provided much insightful information, especially in the area of chemical stimulation. However, our basic knowledge of physical and, especially mechanical stimulation of cells has lagged behind investigations on chemical stimulation. The application of engineering technology and measurement to the problems of mechanical stimulation of soft tissues and the response of cells presents an exciting area for investigation. An ideal system to perform such investigations is the heart. The cellular components of this system have evolved mechanisms to respond to mechanical stimulation. For example, during development and growth of the heart, muscle cells respond to increased pressure and volume by hypertrophic growth.

[1]The work was supported in part by NIH grants HL37669, HL24935 and HL33656.

Although several investigations have reported the effects of mechanical stimulation of cells (3,4), few have addressed the response of the different cells of the heart. We have designed instrumentation that allows for the repeatable mechanical stimulation of cells in either a cyclic or passive nature. The results of these investigations are presented in this report.

MATERIALS AND METHODS

Cell Isolation and Culture.

Cardiac myocytes were isolated from 5 day old rats (3) and plated on silicon membranes (Dow Corning Silastic, 0.01" thick) coated with 20 ug/ml of laminin and allowed to adapt to culture conditions (2). After 3-7 days in culture, the Silastic membranes were placed in an apparatus designed to induce cyclical linear stretch of the substrate at 12 cycles/min with a linear displacement of 10% (2).

Immunofluorescence Microscopy and Cell Measurements.

Control and stretched cells on Silastic substrates were fixed with 2% paraformaldehyde and stained by standard indirect immunofluorescence techniques using antibodies against vimentin or actin (2). Paraformaldehyde fixed myocytes from stretch and control groups were examined by phase contrast microscopy. Length and width measurements of cells from 25 randomly chosen fields were determined using a calibrated ocular micrometer. The length or width of the cells were measured at the longest or widest part of the cell and data expressed in microns.

Synthesis Studies.

Myocytes were incubated in F12K culture medium containing either 3H-uridine (1 uCi/ml) or 14C-tyrosine (0.1 uCi/ml) for 24 hrs either as static controls or cyclic stretch. The cells were washed with medium, precipitated with 10% TCA at 4°C, washed 3 times with 10% TCA, scraped and solubilized in 2% SDS. Part of the sample was analyzed from protein content and the remainder was mixed with

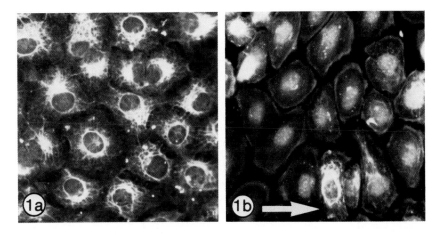

Figure 1.   Immunofluorescence micrographs of endothelial cells grown on silastic membranes and stained for vimentin intermediate filaments.   A) control cells, B) Cells exposed to cyclic linear stretch, the arrow indicates the direction of stretch.

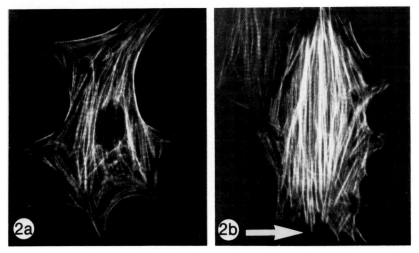

Figure 2.   Immunofluorescence micrographs of myocytes from control (A) and stretched cultures (B).   The arrow indicates the direction of stretch. A) The control cell shows multiple processes and a multidirectional arrangement of the actin filaments, whereas the cell and filaments of the cells exposed to cyclic linear stretch (B) are orientated perpendicular to the direction of stretch.

scintillation   cocktail   and   counted   in   a   liquid
scintillation counter.   The data was calculated as CPM/mg
protein and expressed as a percent increase above control
levels.

RESULTS

When the isolated cells of the heart were subjected to
mechanical stretch of 10% of the substrate, they altered
their orientation from a random pattern to an orientation
that was perpendicular to the direction of the stretch.
The cytoskeleton of the individual cells also showed
changes in orientation.   When endothelial cells are stained
with vimentin, the cytoskeleton forms a basket-like
appearance in the cell.   The staining pattern shows that
the connections of the cytoskeleton extend from the nuclear
membrane to the plasma membrane (Fig. 1). Following
mechanical stimulation, the staining pattern has become
diffuse and lacks the clear connection from the nucleus to
the plasma membrane (Fig. 1).   These data indicate that
the cellular response to the mechanical stretch is a change
in orientation of the cell and rearrangement of the
cytoskeleton into a pattern that allows the cell to be more
resistant to deformation of the mechanical stimulation.
Myocytes and fibroblasts also demonstrated a cellular and
cytoskeletal re-orientation.   Control myocytes (Fig. 2)
exhibited a typical irregular form with multiple processes
containing   contractile   filaments.   When   cyclically
stretched, the cells orientate perpendicular to the
direction of stretch and the contractile apparatus orients
in a similar manner (Fig. 2).   Cells exposed to passive
(static) stretch did not exhibit the changes in cell
orientation and cytoskeleton seen with cyclic stretch.
There was essentially no difference in control or passively
stretched cells at 24 hrs.

MORPHOMETRIC ANALYSIS

	CONTROL	STRETCH	% CHANGE
LENGTH	20.67±0.67	24.57±0.91	18.87
WIDTH	13.13±0.52	15.60±0.52	18.81
LxW	271.39±15.21	383.29±21.97	41.23

Table 1.   Morphometric analysis of neonatal cardiac
myocytes subjected to cyclical linear stretch for 48 hrs.

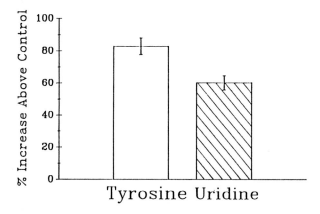

Figure 3. Graphic presentation of the 14C-tryosine and 3H-uridine incorporation into neonatal myoctes that were subjected to cyclical linear stretch for 24 hrs.

In addition to these morphological changes cardiac myocytes underwent what appears to be cellular hypertrophy. These cells showed the typical change in orientation as well as an increase in size (Table 1). The increase in size seemed to be proportional in both length and width. The cells subjected to mechanical stimulation also showed and increase incorporation of 14C-tryosine indicating increased protein synthesis and an increased incorporation of 3H-uridine indicating an increased production of RNA when the cells were mechanically stimulated (Fig. 3). The increase in the area of the cell as well as increased protein synthesis and RNA synthesis strongly suggests that the cells are undergoing hypertrophy in response to mechanical stimulation.

DISCUSSION

The role of mechanical stimulation in relation to cell function has long been speculated but rarely investigated (4,5). Recent studies show that mechanical stimulation can be involved in cell division, wound healing and repair, and other morphogenetic events in tissues and organs. Clearly, mechanical stimulation plays an important role in cardiac function in both health and disease. However until recently it has been difficult to isolate and investigate the effects of mechanical stimulation on the cellular components of the heart (2).

Now that these initial effects have been documented, it is important to investigate the precise events associated with mechanical stimulation (2,6). To do so, we have modified a soft tissue stretching system previously used for retina (6). The application of this technology developed in mechanical engineering can be used to document many of the parameters used in the mechanical stimulation of the cellular components of the heart. Among these, it will be important to measure the amount of deformation of the individual cell when exposed to a known amount of cyclic stretch on the substrate to which it is attached. The stress-strain relationships of normal and stimulated cells will allow for the experimentation on the control of these responses. The relationship of mechanical force to cellular fatigue can also be monitored. Finally, it will be possible to investigate those disease processes which alter the mechanical properties of cells and tissues by being able to understand how the individual cells perceive and response to these forces.

REFERENCES

1.   Leung, D.Y.M., Glagov, S. and Mathews, M.B. Cyclic stretching stimulates synthesis of matrix components by arterial smooth muscle cells in vitro. Science 191:475-477, 1976.

2.   Terracio, L., Miller, B. and Borg, T.K. Effects of cyclic mechanical stimulation of the cellular components of the heart: in vitro. In: Vitro 24:53-58, 1988.

3.   Borg, T.K., Rubin, K.R., Borg, K.L., Lundgren, E. and Obrink, B. Recognition of extracellular matrix components by neonatal and adult cardiac myocytes. Develop. Biol. 104:86-96, 1984.

4.   Curtis, A.S.G. and Seehar, G.H. The control of cell division by tension or diffusion. Nature 274:52-53, 1978.

5.   Beloussov, L.V., Dorfman, J.G. and Cherdantzev, V.G. Mechanical stresses and morphological patterns in amphibian embryos. Embryol. Exp. Morphol. 34:559-574, 1975.

6.   Wu, W., Peters, W.H. and Hammer, M.E. Basic mechanical properties of retina in simple elongation. J. Biomechanical Eng. 109:65-67, 1987.

Tissue Engineering, pages 57–63

REMODELLING OF CARDIAC TISSUE - IS IT POSSIBLE?[1]

Karel Rakusan

Department of Physiology, University of Ottawa
Ottawa, Ontario Canada K1H 8M5

ABSTRACT The close relationship between cardiac
structure and function is discussed. Proliferation
of new cardiac myocytes and growth of coronary
capillaries may play an important therapeutic role.
At the present time, our knowledge of the
regulatory processes for cardiac growth is rather
limited. Possible understanding may be derived
from the changes observed at the tissue level
during normal and pathological growth of the heart,
which will be subsequently described.

INTRODUCTION

Cardiac structure and function are closely
interrelated. For instance, the lack of adequate blood
supply leads to the loss of cardiac myocytes and vessels
as in cardiac infarction. On the other hand,
morphologic features of cardiac hypertrophy (increased
cell size and prolongation of intercapillary distance)
are important contributors to the subsequent failure of
the hypertrophic hearts. Understanding of this
relationship will also lead to a rational design for the
proper treatment or even prevention of some pathological
phenomena.

---

[1]This work was supported by the Ontario Heart and Stroke
Foundation

In the above-mentioned example of cardiac infarction, a possibility of proliferation of additional cells and growth of new vessels would greatly enhance the healing process and thus reduce the resultant extent of damage. Similarly, forming cardiac hyperplasia instead of cardiac hypertrophy, i.e. creating a similar increase in cardiac mass without changing the cell size, together with formation of additional capillaries, would make the heart less prone to cardiac failure. Thus, tissue engineering in this particular case may lead to a significant reduction in the morbidity and, ultimately, the mortality from cardiovascular diseases.

At the present time, our chances of modifying the cardiac tissue are very limited. As a matter of fact, we are not even familiar with the mechanisms which are responsible for the regulation of changes occurring during normal cardiac growth.

NORMAL GROWTH OF CARDIAC TISSUE

Recently (1), we reviewed the basic features of the postnatal development of mammalian hearts, which may be summarized as follows.

Cardiac Myocytes

Early postnatal growth of the mammalian heart is realized by an increase in volume of the individual myocytes as well as in their number. In contrast to older concepts of the heart as a "cell constant" organ, it is now widely accepted that proliferation of muscle cells takes place in newborn hearts. It is not clear, however, when or why the proliferation stops. Linzbach (2) was the first to suggest that hearts from newborns contain only half the total number of myocytes present in the adult human heart (at autopsy). Adult values, however, are rapidly attained during the first few postnatal months. Proliferation of the cardiac myocytes in neonatal canine hearts was suggested by Bishop and Hine (3) and the same has been proposed for rat myocardium by several authors, including ourselves (4). Thus, the production of new muscle cells takes place in the neonatal period, and the capacity of myocytes to proliferate is lost around the weaning period. Subsequent growth of the cardiac mass is accomplished solely by increasing the size of the individual myocytes (hypertrophy), without changing their total number.

Coronary Capillaries

Capillary growth is also strongly influenced by the age of the organism. Early postnatal development is characterized by rapid growth of the capillary network. In the neonatal heart, 4-6 muscle fibers are supplied by a single capillary, this ratio decreases to about 1.5 around the weaning period and 1.0 in the adult. This observation is valid for such divergent species as man, rabbit and rat (1). Recently, we estimated that close to half of all the existing capillaries in the adult rat heart is formed during the first 3-4 postnatal weeks (5). Later on, the capacity of coronary vessels to grow is diminished. In the adult stage, the numbers of both capillaries and myocytes remain constant, and the intercapillary distance increases as a result of increasing cell size. Finally, a disappearance of capillaries has been observed in the senescent rat heart (1).

Developmental stages

Thus, based on the growth of myocytes and capillaries, we may distinguish four developmental stages of the mammalian heart:
1) Neonatal stage, which is characterized by the proliferation of both cardiac myocytes and coronary capillaries; formation of new muscle cells ends before or during the weaning period, while capillaries still continue to grow for some time.
2) Adolescent stage, during which the total number of myocytes becomes constant, but capillaries still continue to grow, albeit at a much slower rate than previously.
3) Adult stage, which is characterized by the growth of the individual myocytes without any change in the number of muscle cells or capillaries.

4)   Senescent stage, in which the variability of cell
size increases, and in which some capillaries may
actually disappear.
    These morphological milestones of cardiac
development have been clearly outlined only recently.
The exact timing of these changes and, even more
importantly, the basic mechanisms responsible are not at
all clear. We still do not know why the cardiac
myocytes and coronary capillaries cease to proliferate
at a certain stage of heart development. Understanding
this phenomenon would help us to answer an even more
important question from a practical point of view: is
it possible to reverse these changes, to stimulate
additional growth of capillaries and myocytes?

PATHOLOGICAL GROWTH OF CARDIAC TISSUE

    The developmental stage of the heart is also the
major determinant of the changes that occur during
pathological growth of the heart. When the heart is
subjected to an increased load (pressure or volume
overload), growth is stimulated and the organ undergoes
major restructuring. Cardiomegaly, introduced early in
development, is characterized by an enhanced multiplica-
tion of myocytes (hyperplasia) together with a
concomitant proliferation of coronary capillaries. On
the other hand, the same stimulus in the adult organism
will result in true hypertrophy, when the total number
of myocytes and capillaries remains unchanged, but the
size of the individual myocytes increases and the
capillary net becomes less dense (1).

POSSIBILITIES FOR CARDIAC RESTRUCTURING

    Is it possible to induce an additional
proliferation of myocytes and capillaries even in the
adult organism? Cardiac hyperplasia as a result of
stimulated cardiac growth in neonates has been described
by several authors; in contrast, the possibility of a
hyperplastic response in the adult heart has been
observed only occasionally and is not generally accepted

(for a review see 6). However, signs of additional
capillary growth in the adult heart have been reported
by several authors in a variety of experimental
situations: in adult rabbits after bradycardial pacing,
and in adult rats with spontaneous hypertension or after
treatment with extracts from fetal heart, propranolol,
dipyridamole, adenosine, or nifedipine (7). In
contrast, an impeded capillary growth has been observed
in neonatal rats chronically injected with protamine
(5).

A relative paucity of data on the growth response
of the cardiac tissue to both normal and pathological
stimuli can probably be explained by the lack of
convenient methods for evaluation of this phenomenon.
Estimates of cell number and evaluations of cardiac
capillarization are time-consuming methods with a
sizeable subjective component. In addition, in the case
of application of various drugs and growth factors, the
exact amount of the substance reaching the cardiac
tissue is not known.

Therefore, many crucial questions concerning the
growth and proliferation of both cardiac myocytes and
coronary capillaries still remain unanswered. These are
questions like:
-    What is the exact age at which mammalian heart
myocytes (from various species) normally stop
proliferating? When does capillary growth stop?
-    Can proliferation be reactivated in the adult by
the introduction of an abnormal or pathological stimulus
for cardiac growth, and if so, for how long can such
growth be maintained?
-    If additional muscle cells and capillaries are
formed, is this increase permanent even after the
disappearance of the growth stimulus?

Both cell proliferation and angiogenesis are at the
present time topics which are being intensively studied.
Most of the studies, however, are at the molecular
level using simplified, often in vitro, models. We are
not aware of any attempts to test any of the proposed
stimulatory substances on intact cardiac tissue. Their
application will depend on two factors: 1) development
of simplified methods for estimation of cell number and
quantitative evaluation of cardiac capillarization, and

2) introduction of an experimental model of cardiac tissue more accessible to application of various growth-stimulating substances. The first task involves streamlining the methods of quantitative morphology and introducing the use of objective cell and tissue analyzers. For the second objective, we propose the utilization of the technique of ventricular grafts implanted in oculo, as used recently by several authors (8,9).

Thus, understanding the mechanisms leading to the growth response of cardiac tissue to normal and pathological growth stimuli should provide us with the key information needed for the long-term goal: the remodelling of cardiac tissue according to the needs of the organism by formation of additional muscle cells and coronary blood vessels.

REFERENCES

1.   Rakusan K (1984) Cardiac growth, maturation and aging. In: R. Zak (ed): "Growth of the Heart in Health and Disease", New York: Raven Press, p.131.
2.   Linzbach AJ (1947) Mikrometrische und histologische Analyse hypertropher menschlicher Herzen. Virchows Arch 326:458.
3.   Bishop SP, Hine P (1975) Cardiac muscle cytoplasmic and nuclear development during canine neonatal growth. In: Roy PE, Harris P (eds): "Recent Advances in Studies on Cardiac Structure and Metabolism. Vol. 8 The Cardiac Sarcoplasm". Baltimore: University Park Press, p.77.
4.   Rakusan K, Jelinek J, Soukupova M, Poupa O (1965) Postnatal development of muscle fibers and capillaries in the rat heart. Phys Bohemoslov 14:32.
5.   Rakusan K, Turek Z (1985) Protamine inhibits capillary formation in growing rat hearts. Circ Res 57:393.

6. Rakusan K, Korecky B, Mezl V (1983) Cardiac hypertrophy and/or hyperplasia? In: Alpert NR (ed): "Perspectives in Cardiovascular Research. Vol. 7 Myocardial Hypertrophy and Failure". New York: Raven Press, p.103.

7. Rakusan K, Turek Z (1986) A new look into the microscope: proliferation and regression of myocardial capillaries. Can J Cardiol 2: 94.

8. Olson L, Seiger A (1976) Beating intraocular hearts: light-controlled rate by autonomic innervation from host iris. J Neurobiol 7:193.

9. Tucker DC, Gist R (1986) Sympathetic innervation alters growth and intrinsic heart rate of fetal rat atria maturing in oculo. Circ Res 59: 534.

Tissue Engineering, pages 65–67
© 1988 Alan R. Liss, Inc.

Summary/Discussion

Soft Tissue Remodeling

Peter D. Richardson

Many early studies on tissue engineering have used macroscopic, continuum representations of mechanical behavior of tissues. In these, some sort of constitutive relationship between stress and deformation is applied to problems for specific tissue geometrics subjected to particular boundary conditions. Different regions within a tissue may have different constitutive relations, reflecting different tissue types. This is useful, for example, in blood vessels, where the tunica intima, media, and adventitia all have different behavior. Some of these representations use strain energy functions, a method of representation which was found useful initially with polymeric materials (of which rubber is the classic example). Strain energy functions have been extended to cover a variety of polymeric materials, even where their mechanical properties are significantly different from rubber. Fracture mechanics, as a topic within continuum mechanics, has made much headway in recent years, but application to tissue damage analysis is yet to be pursued. Even the behavior of muscle in both its activated and passive states has been represented by various authors within the continuum mechanics, large deformation framework of analysis. However, there is ample scope for further development along these lines.

For progress in tissue engineering it is clear that the continuum mechanics representation is insufficient. A bridge to understanding macroscopic behavior in terms of the detailed structure of the tissues is necessary. Features of the detailed structure which need to be considered include the cells, the extracellular matrix, and their interconnection. Some of the methods currently available for study are tedious, and reliable quantification of crosslinking needs further development. Two current sources of interest in understanding the relation between the structure of tissue and its mechanical properties are: (1) Implanted materials elicit tissue reaction, and this can be controlled (in ways still being studied) to provide tissue that carries loads. Bioresorbable implants provide a specific challenge here, because the polymeric scaffold that initially carries the load progressively disappears, leaving the reaction tissue to assume the load-bearing task. (2) Tissues alter as a consequence of age and disease, and it would be useful to assess quantitatively the margin of safety which may remain against development of aneurisms, or local tissue fracture, and information on structure is most likely to be available for this assessment.

Discussion at the meeting drew attention to the interaction between the adaptive modification of tissue structure and the loads imposed upon it. There is the obvious process of adaptation of the size and structure of blood vessels as they progress from embryonic through childhood to adult changes. The cells in the vessel wall are seen as being mechano-receptors, and the production of collagen and its distribution into the structure are presumably continued throughout life. Speculations were made about the significance of different modes of applying loads to the cells: Are hydrostatic stresses more important in determining production and excretion of tropo-collagen, or are shear stresses the more important mode? How do cells sense the different deformations, and are there features in cell membranes and cytoskeletons which mediate the process? What mechanical interactions with the applied loads affect elastin formation and arrangement? May we be able to develop mathematical representations which give useful approximation to the structural modification processes, so that projections can be made about soft tissue development under prescribed conditions?

Part of the discussion on the interaction of vessel structures and the loads imposed upon them was focused on short-term effects, which can be induced and observed in experimental animals. Effects which can be seen very quickly in the vessels of small animals, e.g,. rats, may take much longer to occur to a corresponding degree in larger animals, especially humans. We do not know reliable time-scaling rules in comparing experiments in different species. Experimental conditions discussed included acute induction of hypertension, use of stenosis-generating clamps on the aorta (rats) or on coronary arteries, and autotransplantation of vein segments to arterial positions. Increases seen in production of collagen type I and III in acutely-induced hypertension were cited, and their ratio changed too. "Memory" of short-term mechanical processes that are imposed on soft tissues might be mediated through rearrangement of extracellular matrix around cells.

Y.C. Fung reiterated in discussion the significance of the residual stresses which can exist in tissues, especially arteries. The development, distribution and maintenance of these residual stresses have been largely neglected in the past. It is likely that they have an important function, perhaps in maintaining a certain level of activity in the vessel wall cells that is essential to stability.

Tissue Engineering, pages 69–71
© 1988 Alan R. Liss, Inc.

II. Skin and Connective Tissue

- *Skin*

- *Connective Tissue*

Introduction. . . . . .Merton Bernfield

Skin

A major opportunity for tissue engineering is the replacement of skin, especially the epidermis, in skin loss injuries including burns. Several approaches to obtain epidermal replacements have been attempted to reduce the need for split thickness mesh autografts, the current optimal replacement.

Cultured human keratinocytes, obtained from cadaver donors or skin biopses, will undergo extensive proliferation, producing 102 to 104 - fold more cells than those explanted into culture. These cells will also stratify and differentiate in culture, producing a 97% pure population of keratinocytes in a physically continuous sheet. The sheet can be placed directly onto a suitably debrided, clean wound or onto a composite gel of type I collagen and chondroitin sulfate prior to placement on a wound.The former procedure has been performed on more than 50 allogeneic recipients with success and little apparent immunologic problem.

Introduction of the culture period may allow loss of the immunocompetent cells during expansion of the keratinocyte population. The mechanism by which these cells are tolerated by the recipient is unclear. Also unclear is whether the graft acts as a temporary protective barrier or is incorporated into the newly formed epidermis.

This approach is proving successful, but priority items for study include:
- Standardization of the culture technique. Currently, a variety of media are being used and chemically-defined media would be optimal

● Evaluate the mechanism of graft immunotolerance. Is it due to the loss of immunocompetent cells only, or is there true   loss of transplantation antigens from the keratinocytes. If so, is this loss permanent?

● Determine whether the graft is incorporated into regenerating skin.

● Construct a suitable dermal substitute that allows keratinocyte viability, dermal fibroblast invasion and formation of epidermal adnexae (e.g. hair, sweat glands, sebaceous glands.)

Connective Tissue

The extracellular matrix (ECM) is an insoluble deposit of large, multi-domain molecules that act as a substratum for cells. There are two general types of ECM: (1) interstitial matrix, that lies between connective tissue cells and (2) the basal lamina, that lies beneath parenchymal cells. Although there are differences between these matrices in the specific types of ECM components, each contains: collagen(s), adhesive glycoproteins and protoglycans. The ECM extends over multiple cells and, thus, integrates cellular behavior at the tissue level.

Tissue replacement and tissue repair involves the deposition of cells into new environments that differ substantially from those in situ. These environments will be largely ECM. Moreover, synthetic polymers have been proposed and used as cell substrata both in vitro and in situ. Therefore, efforts must be made to understand the organization of the ECM, how cells respond to ECM,and how synthetic materials can be used to mimic ECM.

Several approaches have been proposed to enhance tissue replacement and repair through the use of ECM materials. These include chemical modification of readily available ECM by processes which render them non-immunogenic and increase their stability in tissues. Major advances have been made in the isolation and purification of adhesive glycoproteins, e.g.fibronectin and laminin, and in their use as factors to promote cell attachment to synthetic materials. Although the ECM of a single tissue is complex, in vitro studies suggest that the presence of a single adhesive allow the deposition by the cell of the proper ECM.

Among the priority items to be considered for enhancing tissue engineering are:

- Define cell adhesion to ECM components. This will require analysis of the matrix receptors on cells, characterization of the interactive domains on the ECM components, and preparation of soluble derivatives that can be used to modify adhesion.

- Define cellular responses to ECM components. This will require knowledge of which ECM components will allow   maintenance of normal differentiated cell function.

- Define the ECM at interfaces between growing and repairing tissues. This will require analysis of extracellular control mechanisms for synthesis and degradation of ECM.

Tissue Engineering, pages 73–79
© 1988 Alan R. Liss, Inc.

THE GRAFTING OF 55 BURN PATIENTS WITH CULTURED ALLOGENEIC EPIDERMIS

John M. Hefton, Ph.D.; Lisa Staiano-Coico, Ph.D.; Michael
R. Madden, M.D.; and Jerome L. Finkelstein, M.D.

Burn Center, The New York Hospital-Cornell Medical Center
New York, NY 10021

We have developed a method for culturing human
epidermal cells from cadaver donors and using these
sheets of cultured cells as coverings on the burn wounds
of non-related, or allogeneic, individuals.(1,2) Using
these techniques we have grafted the burn wounds of 55
patients with cultured allogeneic epidermis. We have
followed some of these patients for more than 5 years
after grafting with cultured allogeneic epidermal cells
and have not observed any episodes of acute rejection.

A major goal in the treatment of burn wounds has
been rapid coverage of the wound to avoid sepsis and
other complications. In order to restore normal skin
function the most appropriate covering has been the
autograft, a split-thickness piece of skin taken from an
unburned area, or donor site, on the body.
Unfortunately, severely burned individuals often have a
limited availability of donor sites for autografting.
Although temporary substitutes for autologous skin, such
as cadaver skin (homograft), pig skin (xenograft), or
synthetic dermal substitutes, have been used; burn
victims eventually require a permanent covering derived
from their own skin. In attempts to increase the amount
of epidermis available for wound coverage many
investigators have used tissue culture derived material.

Over the past 20 years improvements in tissue
culture techniques have produced several methods by
which epidermal cells can be grown in culture as
confluent sheets. These different techniques, however,
yield significantly different types of cultured sheets
with regard to their degree of differentiation and their
ability to influence wound healing. Several groups of
investigators, using different techniques, have grafted
both autologous and allogeneic cultured epidermal cells
onto both animals and human patients.(3,4) The use of
autologous cultured grafts demands the collection of a
split-thickness piece of the patient's skin. Although
autologous cultured epidermal grafts have been
successfully used on human burn wounds this method has
several limitations, including: the limited degree to

which human epidermal cells can be expanded in any
known culture system, the increased time of
hospitalization required by such approaches, and the
relative lack of differentiation demonstrated in
cultures grown according to this techniques.

We have developed a method for culturing epidermal
cells from cadaver donors, and grafting the cultured
cells onto the burn wounds of allogeneic, or non-
related, individuals.(1)   This approach offers several
advantages over an autologous system including: the
ready availability of cultured grafts which can reduce
hospitalization time, the increased epidermal
differentiation which has been demonstrated in cell
grafts produced by these methods, and the fact that
donor sites, which are painful and can become infected,
do not have to be created on the patient.   Epidermal
cells grown according to the method we have developed
differentiate into sheets of stratified cells 10 to 15
cells layers thick (Figure 1).   Epidermal cultures
produced by this technique have been demonstrated to
display many characteristics of normal epidermal
differentiation such as expression of high molecular
weight keratin proteins by upper layer cells, expression
of cell surface differentiation molecules specific for
basal layer and suprabasal cells, and the development of
specific structural organelles such as tonofilaments and
desmosomes.(5)

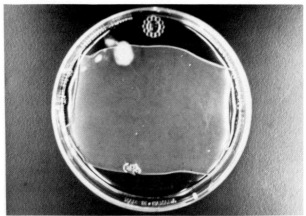

FIGURE 1. A typical cultured epidermal graft after
three weeks growth in vitro and enzymatic removal from
the tissue culture flask.

Melanocytes have also been identified within our cultures by the identification of intracellular melanosomes and melanin. In addition, we have been able to use biochemical agents to induce terminal differentiation in these cultures so that a stratum corneum-like layer of enucleated keratinocytes develops in vitro.(5) Furthermore, we have identified changes in the patterns of distribution and amounts of actin produced in cells which were incubated in the presence of biochemical inducing agents.(5) These cytoarchitectural changes in the cultured keratinocytes have been correlated to the kinetic status of the cultured epidermal grafts at various times of their growth in vitro, and also correlated with the ability of these cultured grafts to promote rapid healing on wounds in vivo.

The growth of human epidermal cells in the culture method we have developed has been extensively characterized by multiparameter flow cytometric analysis. These studies have revealed that cultures of epidermal cells grown according to our techniques are composed of three distinct subpopulations of cells which differ in cell size, RNA content and cell cycle kinetics.(1) These experiments have revealed that there are three distinct phases of culture growth, and that the keratinocyte subpopulations in the cultures undergo significant alterations in kinetics as they progress through these phases. These alterations have been correlated with the phases of wound healing in vivo, and have enabled us to determine the culture age at which the epidermal grafts will produce optimal healing when applied to burn wounds.

We have shown that epidermal cells grown in culture according to our methods do not stimulate the proliferation of allogeneic T lymphocytes in vitro, and do not express HLA-DR transplantation antigens.(6) Further, we have shown that when domestic pigs were grafted with cultures of human epidermis the human tissue could be detected by immunofluorescent assays for up to three months after grafting.(7) Other investigators have employed tissue culture techniques to deplete skin and endocrine tissues of immune-competent cells and have demonstrated that transplants of such tissues survived as functioning allografts.(8)   These results support the hypothesis that epidermal cells grown in culture according to our techniques can have prolonged survival in allogeneic recipients and thus serve as a permanent wound covering.

We have grafted cultured allogeneic epidermis onto the burn wounds of 55 patients (9), and onto the full-thickness stasis ulcers of dermatology patients (10). We have followed some of these patients for more than 5 years after grafting, and not observed any episodes of acute rejection.   However, the patterns of wound healing observed after grafting with cultured epidermis was significantly different between the partial-thickness wounds and the full-thickness wounds. Patients with deep second-degree burn wounds which would usually have required grafting with split-thickness autografts healed in an average of 9 days after being grafted with cultured cells.   The healing of these wounds grafted with cultured cells was of equivalent, or superior, cosmetic quality when compared to the portions of the wounds healed with traditional autografts.   Furthermore, portions of the wounds which were not grafted failed to heal spontaneously and were grafted with either autografts or cultured allografts after the other portions of the wounds were healed.

In contrast, when cultured cells were placed on fat or fascia in full-thickness wounds there was a reduced percentage of take on these wound beds and a slower rate of healing.   These results substantiate the findings of other investigators that the presence of the dermis is important in long-term maintenance of the epidermis.(11) Therefore, for optimal re-epithelialization of skin wounds it appears to be important to transplant cultured epidermal cells in combination with a substitute dermis. The problems involved in this type of approach include the fact that the dermis and the epidermis must each be implanted at the most appropriate time for each in order to ensure that the specialized nutritional demands of each will be met and result in optimal growth and differentiation.

The majority of clinical experiences have shown that the thicker the skin graft, containing both dermis and epidermis, the greater the beneficial effect on

wound contraction and cosmesis (12). Several attempts
have been made to limit wound contraction by the
development of artificial dermal substitutes.(13)
Although the engraftment of such dermal replacements in
full-thickness wounds have decreased the degree of
contraction of the wound during healing, the ultimate
restoration of complete skin function requires an
epidermal covering. The attempts to develop skin
replacement products which can function as a one step
procedure have not succeeded since epidermal function is
compromised during the period of dermal incorporation
and vascularization. In preliminary experiments we have
implanted similar dermal substitutes into full-thickness
wounds in human patients and domestic pigs. We have
observed that the dermal substitutes were sufficiently
vascularized in 7 to 10 days to permit an autograft of
autolgous epidermis or an allograft of cultured
epidermis to survive when placed on the wounds.
Although this is a two step procedure, only the initial
step in which the wound is excised to bleeding tissue
and the dermal replacement implanted needs to be
performed in the operating room. The second step which
involves the placement of the cultured graft on the
vascularized dermal bed can be carried out at the
patient`s bedside without the use of anesthesia. We
have developed procedures for applying cultured
allogeneic epidermal grafts to patients' burn wounds at
the bedside. Our experiences with deep second-degree
wounds which were not excised but covered as described
above revealed complete healing in an average of 11
days, and a significant reduction in the time of
hospitalization for these patients.(9)

    We are currently attempting to define the
mechanisms of wound healing which are involved in the
excellent clinical results we observe. In this regard
we are in the process of documenting the degree to which
the cultured allogeneic cells persist on the wounds
after grafting. We are exploiting the differences in
the HLA genotypes, the HLA phenotypes and the sex
chromosomes of the donor and the recipient tissues to
document this persistence. We have obtained preliminary
evidence that cultured allogeneic epidermal cells placed
on the partial-thickness wounds of two burn patients

persisted on the wounds for up to three months after grafting. Similarly, we have been able to identify human HLA antigen expression on the epidermal cells in biopsies taken from pigs whose partial-thickness wounds had been grafted with cultured human epidermal cell grafts 52 days previously. In addition, in order to maintain a continuous supply of cultured epidermal grafts we have developed methods to store human epidermal cells in a cryopreserved state both before and after growth in culture.

REFERENCES

1.   Staiano-Coico L, Higgins PJ, Darzynkiewicz D, Kimmel M, Gottlieb AB, Pagan-Charry I, Madden MR, Finkelstein JL, Hefton JM. (1986). Human Keratinocyte Culture. Identification and Staging of Epidermal Cell Subpopulations. J Clin Invest 77:396.

2.   Hefton JM, Madden MR, Finkelstein JL, Shires GT. (1983). Grafting of Burn Patients with Allografts of Cultured Epidermal Cells. Lancet ii:428.

3.   Gallico GG, O`Connor NE, Compton CC, Kehinde O, Green H. (1984). Permanent Coverage of Large Burn Wounds with Autologous Cultured Human Epithelium. N Eng J Med 311:488.

4.   Eisinger M, Monden M, Raaf JH, Fortner JG. (1980). Wound coverage by a sheet of epidermal cells grown in vitro from dispersed single cell preparations. Surgery 88:287.

5.   Staiano-Coico L, Hefton JM, Helm RE, Higgins PJ. (1988). Cytoarchitectural changes during human keratinocyte differentiation. J Clin Invest, in press.

6.   Hefton JM, Amberson JB, Biozes DG, Weksler ME. (1984). Loss of HLA-DR Expression by Human Epidermal Cells after Growth in Culture. J Invest Dermatol 83:48.

7.   Hefton JM, Madden MR, Finkelstein JL, Oefelein MG, LaBruna AN, Staiano-Coico L. (1987). The Grafting of Cultured Human Epidermal Cells Onto Full-Thickness Wounds on Pigs. J Burn Care 19:29.

8.   Donohoe JA, Andrus L, Bowen KM, Simenovic C, Prowse
     SJ, Lafferty KJ.   (1983). Cultured Thyroid
     Allografts induce a state of partial tolerance in
     adult recipient mice. Transplantation 36:62.
9.   Madden MR, Finkelstein JL, Staiano-Coico L, Goodwin
     CW, Shires GT, Nolan EE, Hefton JM. (1986).
     Grafting of Cultured Allogeneic Epidermis on
     Second- and Third-degree Burn Wounds on 26
     Patients. J Trauma 26:2505.
10. Hefton JM, Caldwell D, Biozes DG, Balin AK, Carter
     DM. (1986). Grafting of Skin Ulcers with Cultured
     Autologous Epidermal Cells. J Amer Acad Dermatol
     14:399.
11. Briggaman RA, Wheeler CE.   (1968). Epidermal-Dermal
     Interactions in Adult Human Skin Role of Dermis in
     Epidermal Maintenance. J Invest Dermatol 51:454.
12. Peacock EE. (1984). Skin Replacement. In "Wound
     Repair" Third ed., W.R. Saunders Co., pp.218.
13. Yannas IV, Burke JF, Orgill DP, Skrabut EM. (1981).
     Wound Tissue Can Utilize a Polymeric Template to
     Synthesize a Functional Extension of Skin. Science
     215:174.

Tissue Engineering, pages 81–86
© 1988 Alan R. Liss, Inc.

FUNCTIONAL WOUND CLOSURE WITH DERMAL-EPIDERMAL
SKIN SUBSTITUTES PREPARED IN VITRO[1]

Steven Boyce, Ph.D., Tanya Foreman, B.S.,
and John Hansbrough, M.D.

Department of Surgery, H-211; University of California, San
Diego Medical Center; San Diego, California 92103

ABSTRACT   Treatment of full thickness skin loss
injuries, including burns, requires repair of both
the dermal and epidermal components of the skin.
Cultured   human   epidermal   keratinocytes   are
combined with porous and resorbable collagen-
glycosaminoglycan   membranes   to   form   dermal-
epidermal skin substitutes that are organized
histologically like skin. Acceptance of skin
substitutes on athymic mice is confirmed   by
positive staining of healed epidermis for HLA-ABC
antigens. Wound contraction is measured after
treatment   of full-thickness skin wounds with
cultured   skin   substitutes,   human   xenograft,
murine autograft, or no graft. Dermal-epidermal
skin substitutes offer much greater availability
than skin autograft, and can be prepared from
autologous or allogeneic cultured cells. Skin
substitutes with allogeneic cells could be applied
immediately after removal of damaged skin to
minimize patient recovery time and accomplish
permanent wound closure.

INTRODUCTION

Conventional treatment of full-thickness skin loss
injuries, including burns, accomplishes closure of the
wounds with meshed autologous split-thickness skin [1].
Skin autograft meets the criteria for skin substitutes [2].

[1]This work was supported by NIH grant GM35068.

However, autograft donation inflicts injuries at donor sites, autograft expansion ratios have practical limits of about 1:6, and autograft donor sites are not available in large body surface area injuries, such as severe burns. In response to these limitations, a variety of materials have been proposed and tested as substitutes for the epidermal and dermal components of skin autograft [3-8]. Composite skin substitutes consisting of cultured human epidermal keratinocytes (HK) attached biologically to collagen-GAG dermal substitutes may be evaluated by application to full-thickness skin wounds on athymic mice [9]. This report demonstrates the participation of cultured HK of the composite graft in closure of full-thickness wounds on mice.

HK cultures in serum-free media can be expanded rapidly to cover greater than 100 times the area of the original biopsy, and can be cryopreserved in liquid nitrogen [10,11]. Collagen-GAG dermal substitutes are lyophilyzed during preparation and may be stored dry for extended periods of time. Permanent skin substitutes that are prepared in vitro from cultured HK and collagen-GAG substrates [12,13] offer advantages for wound closure that include: a) greater supply for grafting; b) reduction of donor sites; c) fewer surgeries to complete grafting; d) shorter hospitalization time; and e) improved functional and cosmetic results compared to skin autograft. Realization of these advantages requires demonstration in vivo of safety and efficacy that are comparable to skin autograft.

Efficacy of skin substitutes depends, in part, on biocompatibility and uniformity of composition. Preparation in vitro allows control of the biochemical and biologic compositions of skin substitutes, standardized evaluation before application, and modification of composition for specialized purposes. Plasticity of preparative procedures and of composition provides opportunities to modify the cellular, biopolymeric, and biomolecular components of skin substitutes to promote epithelialization and vascularization [14], and simultaneously, to minimize infection and inflammatory responses in the wound. Depletion of immune effector cells (lymphocytes, monocytes, neutrophils) from allogeneic skin by selective culture of HK cells has been demonstrated to result in tolerance of the allogeneic epithelium [15]. True tolerance of allogeneic cultured cells could revolutionize the broad practices of transplantation by providing virtually limitless supplies of tissue and organ substitutes from cryopreserved storage. Principles of biologic plasticity, and of transplantation of

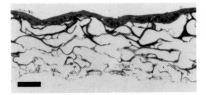

Figure 1.   Light micrographs of HK-collagen-GAG   skin
substitute (left panel),   and of split-thickness human   skin
(right panel).   Scale bars = 0.1 mm.

skin substitutes that contain autologous or allogeneic
cultured cells, can be readily translated to the design and
fabrication of other tissue and organ substitutes.

STRUCTURAL QUALITIES OF SKIN SUBSTITUTES

Structural   qualities of skin substitutes are   compared
histologically   to split-thickness skin in   Figure   1.
Fundamental   considerations   of   structural   composition
include, but are not limited to: a) restriction of the
cultured epithelium to the external surface of the dermal
substitute; and b) total thickness of 0.5 mm or less (Figure
1, left panel).   Lamination of a film of collagen-GAG onto
the surface of the porous collagen-GAG dermal substitute
provides a planar culture substrate. Strict control of
casting procedures and of % wt/vol concentration of starting
materials regulates total thickness of the dermal membrane.
Assuming a constant rate of vascular ingrowth, reduction of
graft thickness results in more rapid vascularization of the
dermal substitute which is required for persistence of the
cultured epithelium. Histological organization of   HK-
collagen-GAG composites compares favorably with split-
thickness skin graft (Figure 1, right panel).

TRANSPLANTATION OF SKIN SUBSTITUTES TO ATHYMIC MICE

Qualitative results of full-thickness wounds after
treatment with dermal-epidermal skin substitutes are shown
in Figure 2.   Healed wounds after treatment with HK-
collagen-GAG composite grafts (Figure 2, left panel) appear
smooth, lightly pigmented and have tensile strength
comparable to surrounding undamaged skin. Wound margins are
identified 10 weeks post surgery by epidermis that stains
positively with a fluorescein labelled monoclonal antibody
against a common hapten of HLA-ABC antigens (Figure 2, right

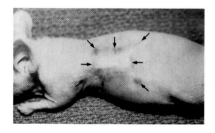 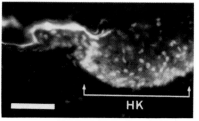

Figure 2.   Wound closure with dermal-epidermal skin
substitutes.   Left panel, perimeter (arrows) of full-
thickness wound on athymic mouse healed after treatment
with skin substitute.   Right panel, net-like pattern of
staining around nucleated epidermal cells identifies human
keratinocytes (HK) at wound margin. Scale bar = 0.1 mm.

panel). Epidermis stained positively for HLA-ABC could only
result from persistence of the cultured HK cells.   These
results demonstrate the participation of cultured HK of the
dermal-epidermal skin substitute in closure of full-
thickness wounds.

Quantitative analysis of wound size six weeks post
surgery is summarized in Table 1.   Wound size, as measured
by planimetry of healed wounds, is used as an index of wound
contraction.   Cultured skin substitutes are intermediate
between split-thickness human skin and no graft in control
of wound contraction.

TABLE 1
% ORIGINAL WOUND SIZE AND STAINING FOR HLA-ABC

TREATMENT	HLA-ABC	% ORIGINAL WOUND SIZE	± SEM
1) Murine Autograft	−	72.83	9.76
2) Human Xenograft	+	46.58	1.83
3) Skin Substitute	+	34.54	4.06
4) No Graft	−	18.42	4.65

DISCUSSION

Emerging technology for preparation of biopolymer
matrices, and of cultured human cells of various types,
provides new capabilties for the design of tissue and organ
substitutes.   However, structural similarity of tissue

substitutes to their native analogues is only the initial
set of considerations for the successful implementation of
tissue substitutes. Graft persistence also depends on
restoration of stable physiologic function and on immune
tolerance of the implant.

Physiologic function, which requires expression of
differentiated phenotype by the cellular components of
the graft, must ultimately be restored after implantation.
In epidermis, a dynamic equilibrium of prolifertive
keratinocyte basal cells, differentiated spinous cells, and
terminally differentiated granular and cornified cells must
be reestablished to restore the protective functions against
infection and fluid loss. Basal keratinocytes retain the
developmental potential to reestablish, in vitro, a
stratified squamous epithelium, but not to regenerate the
epithelial adnexal structures (hair, sweat and sebaceous
glands) of the skin. Formation of cutaneous adnexi has not
yet been demonstrated in skin substitutes prepared in vitro.
This anatomic deficiency of cultured skin results from lack
of instructive (extrinsic) or permissive (intrinsic)
induction [16] by in vitro conditions of adnexal development
that occurs during fetal development. Complete restoration
of anatomic and physiologic function of skin substitutes
will require regulation in vitro of the molecular and
cellular processes which result in the development of skin
adnexi.

Immunologic tolerance of tissue substitutes is required
for initial graft acceptance, for wound healing with
minimum fibrosis, and for long term graft survival.
Tolerance to both cellular and extracellular parts of the
tissue substitute must be achieved. Resorbable polymers,
whether natural (eg., collagen, fibrin) or synthetic (eg.,
polylactide, polyglycolide), are ideally degraded by the
immune system without foreign body reaction, in analogy to
the degradation and repair of damaged tissue in a contusion.
Tissue substitutes with autologous cells offer an optimal
model for study of short-term immune responses during graft
acceptance and wound healing. Although autologous cells in
a graft will minimize the probability of cellular rejection,
they may have limitations of time required for preparation,
or of insufficient availability. Tolerance of implanted
allogeneic cells is hypothesized to result from depletion of
immune effector cells that present high concentrations of
Class II histocompatibility antigens from the donor tissue
[17]. Long-term immune tolerance of grafts with allogeneic
cells could provide unrestricted availability of tissue and

organ substitutes that could be deliberately engineered, cryopreserved and accumulated for replacement of skin, vasculature, endocrine glands, bone, cartilage, nerve and vital organs. Together, principles of developmental biology, tissue culture, polymer chemistry, immunology, wound healing, surgery and other medical disciplines can allow the establishment of new conventions for repair or replacement of transplanted tissues and organs.

REFERENCES

1. Tanner JC, Vandeput J, Olley JF (1964). Plast Reconstr Surg 34:287.
2. Pruitt BA Jr, Levine NS (1984). Arch Surg 119:312.
3. Gallico GG, O'Conner NE, Compton CC, Kehinde O, Green H (1984). New Eng J Med 311(7):448.
4. Burke JF, Yannas IV, Quinby Jr WC, Bondoc CC, Jung WK (1981). Ann Surg 194(4):413.
5. Cuono C, Langdon R, McGuire J (1986). The Lancet 1:1123 May 17, 1986.
6. Bell E, Sher S, Hull B, Merrill C, Rosen S, Chamson A, Asselineau D, Dubertret L, Coulomb B, Lapiere C, Nusgens B, Neveux Y (1983). J Invest Dermatol 81(1), supp:2s.
7. Yannas IV, Burke JF, Gordon PL, Huang C, Rubenstein RH (1980). J Biomed Mat Res 14:107.
8. Boyce ST, Glafkides MC, Foreman TJ, Hansbrough JF (1988). J Burn Care and Rehab, in press.
9. Boyce ST, Sakabu S, Foreman T, Hansbrough J (1987). Program of the 19th annual meeting of the American Burn Association, Abstract No. 78.
10. Boyce ST, Ham RG (1985). J Tiss Cult Meth 9(2):83.
11. Boyce ST, Ham RG (1983). J Invest Dermatol 81(1), supp:33s.
12. Boyce ST, Hansbrough JF (1988). Surgery, in press.
13. Boyce ST, Christianson DJ, Hansbrough JF (1988). J Biomed Mat Res, in press.
14. Irvin TT (1984). In Bucknall and Ellis (eds): "Wound Healing for Surgeons," Philadelphia: Bailliere Tindall, p 3.
15. Thivolet J, Faure M, Demidem A, Mauduit G (1986). Transplantation 42(3):274.
16. Ham RG, Veomett MJ (1980). "Mechanisms of Development." St Louis: CV Mosby, p 406.
17. Lafferty KJ, Cooley MA, Woolnough J (1975). Science 188:259.

Tissue Engineering, pages 87–92
© 1988 Alan R. Liss, Inc.

PLASTICITY OF PHENOTYPIC EXPRESSION IN CELLS OF MESENCHYMAL
ORIGIN: IS KELOID FORMATION DUE TO A DEFECT IN A SWITCH
MECHANISM?[1]

Shirley B. Russell, Joel S. Trupin and James D. Russell

Division of Biomedical Sciences, School of Graduate Studies,
and the Department of Biochemistry, School of Medicine,
Meharry Medical College, Nashville, TN 37208

ABSTRACT   Our laboratory has been studying keloid formation, an inherited disease of wound healing. Our studies are concerned with the influences of hormones and
growth factors on regulation of growth and matrix production in cultured dermal fibroblasts. We have characterized differences between fibroblasts from normal
adult dermis and keloids, and similarities between
keloid and fetal fibroblasts which provide bases for
studying key regulatory pathways in wound healing and
development.

INTRODUCTION

Cells of mesenchymal origin, including fibroblasts,
smooth muscle cells, chondrocytes and osteogenic cells are
characterized by highly regulated programs of growth and
extracellular matrix production. Each cell type expresses
more than one program (1-6). These different programs may
serve different developmental stages of the organism or they
may be reversible patterns that occur in response to an external stimulus such as wounding. Expression of a particular program is mediated by the interaction of target cells
with peptide growth factors and hormones.
Fibroblasts play a central role in skin repair. Initiation of wound healing involves peptide factors from
platelets and inflammatory cells that promote fibroblast
migration into a wound and stimulate proliferation and
matrix production (6). Fibroblast activity may be maintained

[1]Supported by USPHS grants CA17229, DK35284 and RR08037.

by the continued presence of peptides that initiated the response or by exogenous or autocrine factors synthesized is stimulated during wound healing. Termination of wound healing may involve signals that prevent further entry or activity of exocrine factors or that turn off production or activity of autocrine factors.

Keloids are benign collagenous tumors that form during a prolonged wound healing process (reviewed in 7). The genetic predisposition to form keloids occurs with particularly high frequency in some black populations. There is considerable phenotypic heterogeneity even within the same individual; some wounds heal normally, indicating the involvement of environmental factors in the expression of the keloid gene. Like other hereditary tumors, keloids are polyclonal in origin (8,9). Keloids are distinguished from other forms of hypertrophic scarring by their extension beyond the bounds of the wound, failure to respond to treatment, recurrence after surgical removal and failure to regress with time.

Keloid formation is characterized by an extended period of fibroblast proliferation and an elevated rate of collagen synthesis compared to normal wound healing. The rate of collagen synthesis in keloids is high compared to normal dermis or mature scar but is similar to that seen in early wounds (10,11). Histological studies indicate that keloids arise from the reticular layer of the dermis (12).

Identification of the abnormal regulatory mechanism in keloid fibroblasts may help to elucidate normal mechanisms whereby fibroblasts revert to a state of lower synthetic activity.

RESULTS AND DISCUSSION

We have compared parameters of growth and matrix production in fibroblasts derived from normal adult dermis, normal fetal dermis, keloid tissue, normal granulation tissue and mature scar. In tissue culture medium containing 10% fetal bovine serum and ascorbic acid, there are no significant differences in growth or collagen synthesis in fibroblasts cultured from different dermal tissues (13,14). However, when culture medium is supplemented with physiological concentrations of hydrocortisone (1.5 μM), keloid and fetal fibroblasts differ from fibroblasts of normal adult dermis, normal granulation tissue or mature scars. Growth of keloid and fetal fibroblasts is not stimulated by hydrocortisone whereas the maximum density of the other cell types is increased (14). Hydrocortisone reduces the rate of collagen synthesis (14-16) and the levels of mRNAs for types I, III and V collagens in normal fibroblasts from adult dermis and scar, but fails to do so in keloid and fetal fibroblasts (manuscript submitted). The different responses to hydrocortisone are not due to differences in glucocorticoid receptor number, steroid affinity or chromatin binding (18). The abnormal regulation by hydrocortisone in keloid cells is seen only in a subset of glucocorticoid-regulated processes. Glucocorticoid induction of phosphoenolpyruvate carboxykinase (19), noncollagen protein synthesis (14), and fibronectin mRNA (manuscript submitted) is the same in normal and keloid cells.

Table 1

Effect of Hydrocortisone on Growth and Collagen Synthesis in Normal and Keloid Fibroblasts

Tissue	Ratio (Hydrocortisone/Control)		
	DNA	Collagen Synthesis	α_1(I) mRNA
Normal Dermis			
Adult	1.13 ± 0.04 (5)	0.40 ± 0.04 (5)	0.29 ± 0.07 (4)
Fetal	0.83 ± 0.04 (4)	1.01 ± 0.21 (4)	1.23 ± 0.12 (3)
Keloid			
Surface	1.32 ± 0.09 (4)	0.54 ± 0.03 (4)	0.43 (1)
Nodule	0.85 ± 0.06 (7)	0.95 ± 0.20 (7)	1.38 ± 0.07 (4)

Cultures were fed daily with control medium containing 0.28 mM ascorbate or control medium plus 1.5 μM hydrocortisone. Collagen synthesis, collagen mRNA levels and DNA were measured on day 10 after subculture as previously described (14,21). The data are presented as the mean and standard error of replicate assays of the number of different strains indicated in parentheses. The collagen probes were provided by Jeanne Myers.

The hydrocortisone response characteristic of keloid fibroblasts is seen only in fibroblasts cultured from the keloid nodule (Table 1) and persists throughout the lifetime of the cultures (20). Fibroblasts from uninvolved dermis or from superficial layers of keloids behave like fibroblasts from normal individuals (14). The stable expression of this altered phenotype in keloid nodules indicates an epigenetic alteration in a population of fibroblasts involved in wound healing. The similarity in response of keloid and fetal cells suggests that keloids may be inappropriately expressing a regulatory program that is expressed normally during development. A response to hydrocortisone similar to that seen in keloid fibroblasts has been reported in rheumatoid synovial cells (22,23) and in smooth muscle cells cultured from normal rat aorta (24,25). Thus, the response to hydrocortisone exhibited by fetal and keloid fibroblasts may be expressed in a variety of normal cell types under certain circumstances. The abnormality in keloid cells and perhaps in cells of the rheumatoid synovium may be their inability to switch to an alternate regulatory program.

The abnormal behavior of keloid fibroblasts is not limited to their response to hydrocortisone. Whereas growth of adult, fetal and keloid fibroblasts is the same in medium containing 10% serum (Fig. 1B), keloid fibroblasts grow to higher densities than normal adult cells in medium containing 1% serum or 5% plasma (Fig. 1A). Fetal fibroblasts also grow in low serum or plasma (26, Fig. 1A).

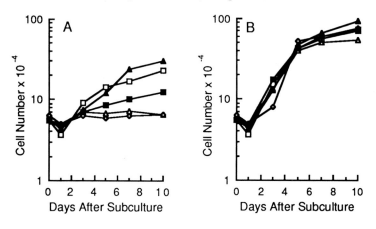

Figure 1.   Growth of Adult, Fetal and Keloid Fibroblasts in Plasma and Serum. Strains derived from normal adult dermis, 131 (Δ) and 103 (✧); normal fetal dermis, 94 (□); and keloid nodules, 50 (▲) and 124 (■), were grown in medium F-10 supplemented with either 5% plasma (A) or 10% serum (B). Cell numbers were determined on the days indicated.

Whereas transforming growth factor ß (TGFß) reduces stimulation of thymidine incorporation by epidermal growth factor (EGF) in normal adult skin and scar, it enhances EGF stimulation in cells from keloids (26). It has been reported that, depending on fetal age, TGFß can enhance or inhibit the response of fetal fibroblasts to other growth factors including EGF (27). Although TGFß affects a growth parameter differently in normal adult and keloid cells, it stimulates the apparent rate of collagen and non-collagen protein synthesis to the same extent in both cell types (unpublished data). The abnormal effects of hydrocortisone and TGFß on certain aspects of keloid cell activity are likely to be due to a single lesion in the complex network through which these and other molecules regulate growth and collagen synthesis.

We have demonstrated that cultured fibroblasts retain complex regulatory patterns that reflect wound healing behavior. The choice of the pattern expressed is influenced by the single gene mutation that predisposes to keloid formation and by environmental factors that affect its expression. The cultured cell system may be used to characterize alternative regulatory patterns and to identify the mechanism that governs the switch from one pattern to another. Such information may be essential in bioengineering of connective tissue.

ACKNOWLEDGMENTS

The expert technical assistance of Kathryn M. Trupin, Alan H. Broquist and Joan C. Smith is gratefully acknowledged.

REFERENCES

1. Duance VC, Bailey AJ (1981). Biosynthesis and degradation of collagen. In Glynn LE (ed) "Tissue Repair and Regeneration," New York: Elsevier, p 51.
2. Raghow R, Kang AH, Pidikiti D (1987). Phenotypic plasticity of extracellular matrix gene expression in cultured hamster lung fibroblasts. Regulation of type I procollagen and fibronectin synthesis. J Biol Chem 262:8409.
3. Ross R (1986). The pathogenesis of atherosclerosis - an update. New Eng J Med 314:488.
4. Hay ED (1981). Collagen and embryonic development. In Hay ED (ed): "Cell Biology of Extracellular Matrix," New York: Plenum Press, p 379.
5. Caplan AI (1987). Bone development and repair. BioEssays 6:171.
6. Castor CW (1981). Autacoid regulation of wound healing. In Glynn LE (ed): "Tissue Repair and Regeneration," New York: Elsevier, p 177.
7. Murray JC, Pollack SV, Pinnell SR (1981). Keloids: a review. J Am Acad Dermatol 4:461.
8. Trupin JS, Williams JM, Hammons J, Russell JD (1977). Multicellular origin of keloids. Fifth International Conference on Birth Defects, Montreal Canada, Abstract.
9. Moulton-Levy P, Jackson CE, Levy HG, Fialkow PJ (1984). Multiple cell origin of traumatically induced keloids. J Am Acad Dermatol 10:986.
10. Cohen IK, Diegelmann RF, Keiser HR (1976). Collagen metabolism in keloid and hypertrophic scar. In Longacre JJ (ed) "The Ultrastructure of Collagen," Springfield: Charles C. Thomas, p 199.
11. Craig RDP, Schofield JD, Jackson DS (1975). Collagen biosynthesis in normal and hypertrophic scars and keloid as a function of the duration of the scar. Br J Surg 62:741.
12. Garb J, Stone MJ (1942). Keloids: review of the literature and a report of eighty cases. Am J Surg 43:315.

13. Russell JD, Witt WS (1976). Cell size and growth characteristics of cultured fibroblasts isolated from normal and keloid tissue. Plast Reconstr Surg 57:207.
14. Russell JD, Russell SB, Trupin KM (1978). Differential effects of hydrocortisone on both growth and collagen metabolism of human fibroblasts from normal and keloid tissue. J Cell Physiol 97:221.
15. Russell SB, Russell JD, Trupin KM (1981). Collagen synthesis in human fibroblasts: effects of ascorbic acid and regulation by hydrocortisone. J Cell Physiol 109:121.
16. Trupin JS, Russell SB, Russell JD (1983). Variation in prolyl hydroxylase activity of keloid-derived and normal human fibroblasts in response to hydrocortisone and ascorbic acid. Collagen Rel Res 3:13.
17. Russell JD, Russell SB, Trupin KM (1982). Fibroblast heterogeneity in glucocorticoid regulation of collagen metabolism: genetic or epigenetic? In Vitro 18:557.
18. Gadson PF, Russell JD, Russell SB (1984). Glucocorticoid receptors in human fibroblasts derived from normal dermis and keloid tissue. J Biol Chem 259:11236.
19. Arinze IJ, Raghunathan R, Russell JD (1978). Induction of mitochondrial phosphoenolpyruvate carboxykinase in cultured human fibroblasts. Biochim Biophys Acta 521:792.
20. Gayden A, Russell SB (1985). Growth and collagen metabolism in senescing human diploid fibroblasts. Fed Proc 44:1063.
21. Abergel RP, Pizzurro D, Meeker CA, Lask G, Matsuoka LY, Minor RR, Chu M-L, Uitto J (1985). Biochemical composition of the connective tissue in keloids and analysis of collagen metabolism in keloid fibroblast cultures. J Invest Dermatol 84:384.
22. Castor WC, Dorstewitz EL (1966). Abnormalities of connective tissue cells cultured from patients with rheumatoid arthritis. I. Relative unresponsiveness of rheumatoid synovial cells to hydrocortisone. J Lab Clin Med 68:300.
23. Castor WC (1971). Abnormalities of connective tissue cells cultured from patients with rheumatoid arthritis. II. Defective regulation of hyaluronate and collagen formation. J Lab Clin Med 77:65.
24. Longenecker JP, Kilty LA, Johnson LK (1984). Glucocorticoid inhibition of vascular smooth muscle cell proliferation: influence of homologous extracellular matrix and serum mitogens. J Cell Biol 98:534.
25. Leitman DC, Benson SC, Johnson LK (1984). Glucocorticoids stimulate collagen and noncollagen protein synthesis in cultured vascular smooth muscle cells. J Cell Biol 98:541.
26. Russell SB, Trupin KM, Rodriguez-Eaton S, Russell JD, Trupin JS (1988). Reduced growth-factor requirement of keloid-derived fibroblasts may account for tumor growth. Proc Natl Acad Sci USA 85:587.
27. Hill DJ, Strain AJ, Elstow SF, Swenne I, Milner RDG (1986). Bifunctional action of transforming growth factor-ß on DNA synthesis in early passage human fetal fibroblasts. J Cell Physiol 128:322.

Tissue Engineering, pages 93–95
© 1988 Alan R. Liss, Inc.

Summary/Discussion

Skin and Connective Tissue

Katherine H. Sprugel

The ideal "artificial" skin should have the functional characteristics of skin: the barrier, sensory, immune and cosmetic properties of the epidermis, and the flexibility, vascular supply and structural characteristics of the dermis. In addition, Tavis and colleagues (1) have described a series of properties important in a skin substitute suitable for clinical use: the ability to adhere to the wound bed, transport water vapor, provide elasticity and durability, act as a barrier to bacterial invasion; and be nontoxic, nonantigenic, antiseptic, hemostatic, easy to apply and cost effective. The papers presented in this session represent two of the major approaches to the design of permanent skin substitutes. The first paper described the cultivation and use of cultured human epidermal cells for skin grafting, while the second presented a bilayer skin substitute consisting of an acellular connective tissue sponge covered by a sheet of cultured human epidermal cells. The discussion focused on current limitations of these approaches and ideas for improving them.

The use of cultured epidermal cells in skin replacement presents exciting prospects for therapy but still has some potential problems. The major issues identified were: (1) immunogenicity of donor cells, (2) maturation, and (3) accessory functions. Current culture techniques appear to reduce or eliminate the propagation of most immunoreactive cells, allowing the use of allogeneic cells for grafts. Preliminary clinical experiences are promising, but concern remains with respect to long-term effects. A second issue is the ability of the cultured keratinocytes to mature under cell culture conditions and to what degree this is necessary prior to grafting. In Dr John M. Hefton's studies, the keratinocytes differentiate extensively in culture, including the development of the membrane-coated granules associated with the barrier functions of skin, as well as other markers of epithelial differentiation. The third issue is an area for future work — restoring some of the accessory cells and/or functions to regenerating skin to improve it functionally and cosmetically. This would include structures such as sebaceous glands, hair follicles, melanocytes, Langerhans cells, and sensory innervation.

Cultured epidermal sheets do not work well on third degree burns and other full thickness skin deficits, probably because they do not facilitate regeneration of dermis. Bilayer designs of skin replacements which provide a scaffolding for development of a neodermis appear to have better prospects for full thickness skin wounds. A limiting step in the use of the bilayer approach is replacement of the artificial dermis scaffolding with a neodermis, and concomitantly, maintenance of the epidermal layer while blood vessels to provide nutrition and support functions move into the neodermis. Future approaches to address this issue include: (1) incorporation of appropriate growth factors to stimulate cell migration, proliferation and vascularization in the neodermis, (2) incorporation of antibiotics to minimize infection, and (3) direct seeding of the artificial dermis with cells.

Composition of the connective tissue matrix used in skin replacements may be critical. In the experiments reported by Dr.Steven T. Boyce, a lyophilized bovine collagen/chondroitin 6-sulfate mixture was used as a support for cultured epidermal cell sheets. Chondroitin 6-sulfate is readily available and appears to reduce the thrombogenicity of admixed collagen. Given the known heterogenity in connective tissue components in different sites in the skin and perhaps even in different levels at the same site, it may be productive to test other matrix components and more complex mixtures to optimize integration of the skin replacement into the tissue. Three factors are of concern in designing an optimal matrix: (1) resorption characteristics, (2) ability to support or induce vascularization and cell colonization, and (3) effects of matrix density, thickness and components on wound contraction.

A totally different approach to tissue replacement which might benefit from improved engineering is the process of *in situ* tissue expansion. By inserting an inflatable balloon under the normal skin adjacent to a graft site and inflating it gradually, it is possible to induce intact skin to expand its size by stretching and growing, producing tissue suitable for use as a skin flap or autograft. This approach can provide autologous tissue with appropriate pigmentation and hair density, and when used as a skin flap, also has innervation. The methods in use at this time are relatively crude and the mechanisms poorly understood. Improvements in the devices used to expand the tissue and innovation in the delivery of pharmacologic agents to the target tissue could be quite fruitful.

Reference

1. Tavis, M.J., Thornton, J.W., Danet, R., and Bartlett, R.H. Current status of skin substitutes. Surg. Clin. North Amer. **58:**1233-1248 (1978).

Tissue Engineering, pages 97–98
© 1988 Alan R. Liss, Inc.

III. Implants, Tissue Responses to Implants and Biomaterial Centered Infections

Introduction. . . . . C. Fred Fox

Tissue Engineering encompasses a wide range of implantation approaches. These employ both living (transplantation) and non-living materials. The latter class includes both biomaterial-derived and totally synthetic substances, and this section treats just a few examples of such nonliving implants.

Nonliving implants have a relatively long and successful history of application in the area of orthopaedic devices. Though these devices are nonliving, they are not inert, and induce a wide variety of tissue responses. They may also serve as habitats for microorganisms which evolve cellular exudates that protect invading pathogens from the immune system. Persistent biomaterial-centered infections are a major cause of device rejection, and the serious consequences of compatibility with pathogens must be considered in choosing material for implant manufacture. The papers by Gristina and Hatcher and their coworkers treat selected examples of this problem, as well as approaches to finding solutions.

Most implants make contact with a variety of cell types. One particularly well studied and successful implant of this sort is percutaneous, i.e., coexisting in contact with both living tissue and the external environment. Percutaneous implants, in spite of the apparent opportunities they create for pathogen entry, have enjoyed substantial long term success in dental replacement therapy. Bjursten and Squier and their associates have reviewed many of the factors that must be considered in studies aimed at predicting the long term behavior of a percutaneous device.These include considerations of the active roles of device materials as inducers of tissue differentiation which may or may not support long term functional compatiblity between device and host.

Biomaterial-derived but nonliving implants are no less inert. Nimni and his colleagues describe short term reactions of living tissues to demineralized collagenous implants, some of which can have chondrogenic and/or osteogenic capacities. Bioresponses of this sort can be modified by treating implant materials with site-specific chemical reagents.

The final paper in this section, by Bicher, considers short range metabolic interactions which are seldom considered systematically in studies on implant behavior. This paper underscores the rudimentary nature of the state of knowledge in this area, and also the wide range of opportunity for application of engineering principles to predict long term behavior of new implant materials.

Tissue Engineering, pages 99–107
© 1988 Alan R. Liss, Inc.

BIOMATERIAL SURFACES: TISSSUE CELLS AND BACTERIA,
COMPATIBILITY VERSUS INFECTION

Anthony G. Gristina,[1] Quentin N. Myrvik[2]
and Lawrence X. Webb[1]

Wake Forest University Medical Center
Winston-Salem, North Carolina 27103

ABSTRACT   The major barriers to the extended use of
biomaterials are bacterial adhesion to surfaces, which
causes infection, and unsuccessful tissue integration
or compatibility. Interactions of biomaterials with
bacteria and tissue cells are directed by specific
receptors and outer membrane molecules on the cell
surface and by the atomic geometry and electronic state
of the biomaterial surface. Modifications to biomaterial surfaces at an atomic level will allow the pro-
gramming of cell-to-substratum surface events.

INTRODUCTION

The application of biomaterials as tissue or organ
replacement devices and as components of bioartificial organs
has highlighted biomaterial and adjacent tissue system
interactions.
Basically "inert" biomaterials have been developed,
however, physical and chemical inertness is relative. At an
atomic level there is always some degree of elemental interaction. In short, all biomaterial implants create a pheno-
menological interface at the implant-to-cell system junction
which ultimately requires characterization and modification.

[1]Present address: Section of Orthopedic Surgery, Wake
Forest University Medical Center, 300 South Hawthorne Road,
Winston-Salem, NC 27103
[2]Present address: Department of Microbiology and
Immunology, Wake Forest University Medical Center, 300 South
Hawthorne Road, Winston-Salem, NC 27103

BIOMATERIAL MORBIDITY

The increasing application of biomaterials as temporary
and permanent implants or components of artificial organ
devices has called attention to the most destructive compli-
cation of their use, "biomaterial-centered infection."
Worldwide, millions of patients require tissue and organ
replacement or the use of biomaterial monitoring or drug
delivery devices. Infections occur in less than 1% of total
hip replacements, but in 2% to 4% of total knee replacements
and in 7% of total elbow replacements. Extended-wear contact
lenses may cause bacterial keratitis. Vascular grafts become
infected in 6% of specific risk groups. Intravascular catheters almost always become infected if not changed at regular
intervals. The Total Artifical Heart (TAH) is at risk for
infection 100% of the time if left in place for more than one
month (1). Ventricular assist devices develop infections
20% of the time in use under 31 days (2). The morbidity and
cost of infection associated with failed devices is a significant problem. There is no satisfactory revision for ampu-
tation or death, which is the end result of 30% of infected
aortofemoral prostheses. When an artificial heart becomes
infected, it must be removed and replaced by a transplant,
which may also become infected.

The pathogenesis of biomaterial-centered infection is
explained by adhesive polymicrobial bacterial colonization
of implant substrata which, as inanimate structures, resemble
surfaces in nature for which bacteria have developed adhesive
colonization strategies. Healthy tissue surfaces in hosts
with a competent defense system are resistant to colonization, but once tissues are damaged by disease, trauma, or
surgery, they become susceptible substrata.

MICROBIAL ADHESION AND TISSUE INTEGRATION

Biomaterial implants are usually not well integrated
(colonized) by healthy tissue cells. Biomaterial surfaces
are essentially unsatisfied or unoccupied except by matrix
proteins which are excellent ligands for bacterial receptors.

Microbial adhesion, aggregation and disaggregation
(dispersion) involve interactions between cells and substrata
surfaces in an ambient fluid milieu. The surface and environment may be any that support life (3,4).

In biomaterial infections, compromised tissue or
prosthetic devices provide those substrata. Proteinaceous

adhesins, polysaccharide polymers, and surface and milieu
substances interact and intermix to form an aggregate of

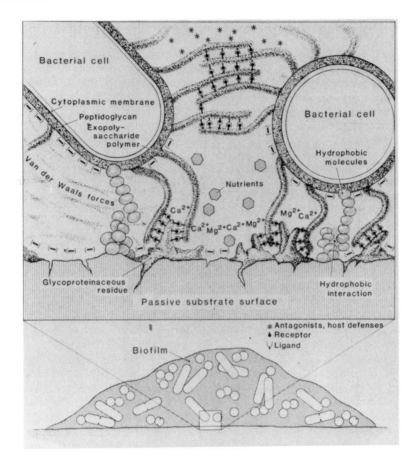

FIGURE 1. Mechanism of bacterial adherence. At specific distances the initial repelling forces between like
charges on the surfaces of bacteria and substrate are overcome by attracting van der Waals forces, and there are
hydrophobic interactions between molecules. Under appropriate conditions there is extensive development of exopoly-
saccharide polymers, allowing ligand-receptor interaction
and pertinaceous binding of the bacteria to the substrate.
Reprinted with permission from A.G. Gristina, et al. Adherent
bacterial colonization in the pathogenesis of osteomyelitis.
Science 1985;228:990-993.

bacteria, elemental substances, glycoproteins, and polysaccharides in a biofilm (Figure 1) (3,4).

Additional symbiotic species may join and present as a
polymicrobial infection. Characteristically, these infections do not respond to treatment until the substratum is
removed. Bacterial adhesion, therefore, and its features
(aggregation to other organisms and the protective nature of
extracapsular slime) direct the pathogensis of damaged tissue
and biomaterial infections (3,4).

Tissue integration is desired for biocompatibility of
certain biomaterials. Integration requires a form of eukary-
otic adhesion or compatibility and possible chemical interaction with an implant surface (5,6). For hemodynamic
systems, however, a biocompatible, nonadhesive, luminal
surface is desirable to prevent thrombus formation or
infection.

THE "RACE FOR THE SURFACE"
(Hypothesis)

As a biomaterial is implanted in a host, it presents an
essentially naked surface of defined atomic structure which
is immediately available to and must interact with milieu
cells, macromolecules and ions. The fate of a surface may
be conceptualized as a "race for the surface" involving
macromolecules, bacteria, and tissue cells. The critical
issue of the cellular response to these surfaces is determined not only by the reactive surface of the cell but by
the atomic and electronic state of the biomaterial outer
layers (up to $2\,\mu$) at the instant of implantation. Observations based on the use of implanted biomaterials in human
hosts suggest that adhesive or integrative phenomena for bacteria or tissue cells and substrata surfaces are critical,
interrelated and based on similar molecular mechanisms.

At implantation, biomaterial surfaces with available
unsatisfied bonds and potential receptor sites for bacteria
or tissue represent domains for potential colonization. Free
energy sites encourage molecular and possibly catalytic
activity and satisfaction by available elements, macromolecules, or cells. All surfaces, regardless of preparation,
acquire a sequence of organic and ionic contaminants whose
distribution is directed by specificities of the implant outer
atomic layers. When tissue cells colonize a metal or polymer
surface and integration with the surface is established,
either via direct chemical interaction or via host-derived

macromolecules, then subsequently arriving bacterial cells are confronted by a living, integrated tissue surface. If not traumatized or altered, this integrated surface is basically resistant to bacterial colonization by virtue of its viability, intact cell membranes, exopolysaccharides and functioning host defense mechanisms. In vivo, bacteria, which are primitive and have a greater ability to colonize nonliving surfaces, may defeat host tissue cells in "the race," causing infection, instead of allowing tissue integration. When bacterial adhesion has occurred, and a stabilized microcolony has developed, it is unlikely that tissue cells will be able to displace the primary colonizers to occupy and integrate the surface. Most biomaterials, and especially complex organ devices, are in part susceptible to infection because at the present state of the art, they are not well integrated. Their surface-tissue interface zone appearance may be characterized at an ultrastructural level as a foreign body reaction.

The race for the surface, therefore, is a contest between tissue cell-to-surface integration of, and bacterial adhesion to, the biomaterial surface. Host defense responses are a major and determinative factor, especially as they are perturbed by biomaterials.

Conditioning Films

Glycoproteinaceous conditioning films, derived from fluid or matrix phases containing fibronectin, fibrinogen, collagen, and other proteins, immediately coat a biomaterial or tissue implant and act as receptor sites for bacterial or tissue adhesion (7). The role of each constituent of this layer differs for each bacteria or type of tissue cell. The three-dimensional mosaic of protein deposition and layering is ultimately directed by biomaterial elemental arrangements and energy hierarchies. Staphylococcus aureus has discrete binding sites for collagen and fibronectin (3,8).

Bacteria in Biomaterial Sepsis

The variety of organisms recovered from infected biomaterials is increasing as methodology improves. S. epidermidis is frequently involved when the biomaterial surface is a polymer or when a polymer is a component of a complex device and is a leading cause of infections of vascular prostheses,

neurosurgical shunts, orthopedic implants, the TAH, and
extended-wear contact lenses (9,10). S. aureus is most often
associated with biometal, bone and joint, and soft tissue
infections and is the most common pathogen isolated in osteomyelitis (11). Pseudomonas aeruginosa is the most frequent
cause of bacterial keratitis associated with the use of
extended-wear contact lenses (12) and is a prevalent pathogen
in TAH cases. Polymicrobial infections appear to be an important feature of substratum-induced infections.

Host defense response is clearly the least studied and
the most critical issue in biomaterial and tissue substitution systems. There are indications that biomaterials
diminish normal host defense mechanisms by interfering with
the phagocytic response.

SUBSTRATA FOR BACTERIAL AND TISSUE CELL ADHESION

The principal biometals in contemporary use are pure
titanium, titanium aluminum vanadium alloy (Ti6Al4V), a
chrome-cobalt based alloy (Vitallium), and stainless steel.
Surface oxides form spontaneously or are created during
production by accelerated nitric acid passivation. Surface
oxides are the reactive interface with host-derived ions and
glycoproteinaceous molecules of the conditioning film and
possibly directly with the surfaces of bacterial cells.

Polymers in common use are polymethylmethacrylate
(Acrylic^R), ultra-high molecular weight polyethylene (Hyfax^R),
polytetrafluoroethylene (Gortex^R), polyurethane (Biomer^R),
and polyethylene terephtalate (Dacron^R). Adsorbates of the
surface of solid polymers tend to satisfy the residual
binding capacity, resulting in decreased surface energies (13).
The hierarchies that result are not as complex as those of
higher energy surfaces such as metals or ceramics.

Tissue adhesion to polymers such as methylmethacrylate
and polyurethane is often characterized by an inflammatory
interface, especially after wear (6). High hydrophobicity
polymers are adhesive for many bacterial species, including
pathogens (7).

Traumatized soft tissue and bone is represented by
amorphous organic fragments of cellular tissue and matrixes,
is rich in microbial nutrient material and ligands, and provides a surface for colonization by bacteria that possess the
appropriate receptors or adhesive polysaccharides. Inanimate,
passive, and fertile damaged tissues are unable to resist
colonization.

Endothelial cells are surrounded by a well-developed
extracellular glycocalyx. When this outer margin is traumatized, receptor sites and fibronectin may be exposed (14,15).
Fibronectin may then be available for bacterial adhesion
(8,16).

PROBLEMS OF COMPLEX ORGAN AND VASCULAR DEVICES

The TAH is a composite of many materials and involves
compatibility among the materials and between materials and
adjacent tissues. Added complexity is created by the need
for adhesive (solid system tissue integration) and antiad-
hesive (fluid environment or hemodynamic system compatibility) surfaces. A drive line traverses organ space, body
cavities, and skin to the external environment, serving as a
pathway for the power source and for microbes. Each biomaterial surface of the TAH favors a particular colonizing
species. The hydraulic interactions required in the device
create eddies and tissue damage that may initiate clotting
cascades and the initial events of microbial adhesion. The
surgical joining of synthetic vessel and natural vasculature
creates a site of intimal perturbation, inflammation, and
endothelial damage, exposing potential receptor sites for
bacterial adhesion (17). Vascular shear forces are sufficient
to dislodge septic or thrombotic aggregates that may have
accumulated on luminal biomaterial or damaged tissue surfaces.
Some of the same critical phenomena may also cause failure
of simple vascular grafts. Sepsis at a vascular graft site
has particularly morbid consequences.

THE FUTURE

Antibiotics impregnated into the surface or bulk of
biomaterials are probably indicated at the present state of
the art. The basis of prophylaxis is that the antibiotics
are in the ecosystem before bacteria can adhere and develop
protective biofilms. Nonadherent bacteria are far more
susceptible to antibiotics. Blocking or saturating analogs
should also be effective. Specific pathogens can be predicted for each material and tissue system, therefore,
correct preventive antibiotics can be chosen. Precoloniza-
tion by healthy tissue cells (osteoblasts, fibroblasts, endothelial cells) may also protect against infection.
Insights in surface atomic and molecular phenomena will

be forthcoming based on the use of advanced instrumentation
such as scanning tunneling microscopy, Auger electron spectroscopy, and electron spectroscopy for chemical analysis,
among others. Surface modification may be the key to controlled biologic responses.

In the future, surface modification techniques will be
used to create idealized surfaces with programmed surface
quantum states or energy levels, suggesting diminished
adhesion for organic or ionic moieties which are present in
conditioning films or on adhesive bacterial cell surfaces.
Alternatively, adhesive zones or margins may be created for
desired biocompatibility and/or tissue integration or seeding.
Methods such as heavy ion implantation, chemical vapor deposition, and vacuum evaporation may be used to create a
surface which "directs" tissue or desired macromolecular
integration for tissue and hemodynamic systems, rather than
bacterial adhesion.

REFERENCES

1.   Artificial Heart Issue (1988). JAMA 259:849.
2.   Farrar DJ, Hill JD, Gray LD, Pennington G, McBride LR,
      Pierce WS, Pae WE, Glenville B, Ross D, Galbraith TA,
      and Zumbro GL (1988). Heterotopic prosthetic ventri-
      cles as a bridge to cardiac transplantation. N Engl
      J Med 318:333.
3.   Christensen GD, Simpson WA, Beachey EH (1985). Adhe-
      sion of bacteria to animal tissues: complex mechanisms.
      In Savage DC, Fletcher M (eds): "Bacterial Adhesion:
      Mechanisms and Physiological Significance," New York:
      Plenum Press, p 279.
4.   Savage DC, Fletcher M (eds) (1985). "Bacterial
      Adhesion: Mechanisms and Physiological Significance,"
      New York: Plenum Press.
5.   Kasemo B, Lausmaa J (1986). Surface science aspects on
      inorganic biomaterials. CRC Crit Rev Biocompat 2:335.
6.   Albrektsson T (1985). The response of bone to titanium
      implants. CRC Crit Rev Biocompat 1:53.
7.   Dankert J, Hogt AH, Feijen J (1986). Biomedical poly-
      mers: bacterial adhesion, colonization, and infection.
      CRC Crit Rev Biocompat 2:219.
8.   Switalski LM, Ryden C, Rubin K, Ljungh A, Hook M,
      Wadstrom T (1983). Binding of fibronectin to
      Staphylococcus strains. Infect Immun 42:628.

9. Bandyk DF, Berni GA, Thiele BL, Towne JB (1984). Aortofemoral graft infection due to Staphylococcus epidermidis. Arch Surg 119:102.

10. Sugarman B, Young EJ (eds) (1984). "Infections Associated with Prosthetic Devices." Boca Raton, Florida: CRC Press.

11. Gristina AG, Oga M, Webb LX, Hobgood CD (1985). Adherent bacterial colonization in the pathogenesis of osteomeylitis. Science 228:990.

12. Slusher MM, Myrvik QN, Lewis JC, Gristina AG (1987). Extended-wear lenses, biofilm, and bacterial adhesion. Arch Ophthalmol 105:110.

13. Andrade JD, Gregonis DE, Smith LM (1985). Polymer surface dynamics. In Andrade JD (ed): "Surface and Interfacial Aspects of Biomedical Polymers, vol. 1, Surface Chemistry and Physics." New York: Plenum Press, p 15.

14. Birinyi LK, Douville EC, Lewis SA, Bjornson HS, Kempczinski RF (1987). Increased resistance to bacteremic graft infection after endothelial cell seeding. J Vasc Surg 5:193.

15. Hamill RJ, Vann JM, Proctor RA (1986). Phagocytosis of Staphylococcus aureus by cultured bovine aortic endothelial cells: model for postadherence events in endovascular infections. Infect Immun 54:833.

16. Webb LX, Myers RT, Cordell AR, Hobgood CD, Costerton JW, Gristina AG (1986). Inhibition of bacterial adhesion by antibacterial surface pretreatment of vascular prostheses. J Vasc Surg 4:16.

17. Gristina AG, Dobbins J, Giammara B, Lewis JC, DeVries WC (1988). Biomaterial-centered sepsis and the Total Artificial Heart. JAMA 259:870.

Tissue Engineering, pages 109–114
© 1988 Alan R. Liss, Inc.

STAPHYLOCOCCUS AUREUS INFECTION OF HUMAN ENDOTHELIAL CELLS[1]

F.D. Lowy, E.A. Blumberg,[2] D.C. Tompkins, D. Patel, V. Bengualid, A. Adimora,[3] and V.B. Hatcher

Department of Medicine and Biochemistry, Montefiore Medical Center, Albert Einstein College of Medicine, Bronx, NY 10467

ABSTRACT Staphylococcus aureus is recognized as a virulent pathogen which is notable for its ability to colonize and invade endovascular tissue. To study the interaction of staphylococci with endothelium, we have developed an in vitro infection assay using cultured human endothelial cells. We have demonstrated that infection involves a sequence of steps including adhesion, phagocytosis and intracellular replication which is followed by cell injury or lysis. Susceptibility of the endothelial cell to infection can be altered by biochemical modulation of these cells with acidic fibroblast growth factor or heparitinase. Following infection, modulation of the endothelial cells occurred. There was an increase in Fc receptor expression detected. Thus, endothelial cell modulation can occur prior to and following infection of the endothelial cells.

INTRODUCTION

Staphylococcus aureus is recognized as a virulent pathogen, which continues to cause life threatening disease. A notable feature of this species is its capacity to colonize and invade endovascular tissue. As a result, S. aureus bacteremia is often complicated

---

[1]This work was supported by grant HL34171 from the National Institutes of Health and a Grant-in-Aid from the American Heart Association. EAB, DCT and AA were supported by NIH training grant AI07183.
[2]Present Address: Department of Medicine, Hahnemann University
[3]Present Address: Department of Medicine, Harlem Hosp

by the development of metatastic suppurative collections
or acute infective endocarditis (1-2). We have utilized
an in vitro infection assay with confluent monolayers
of human endothelial cells to study the sequence of
events associated with staphylococcal invasion. In the
present discussion, we describe some of the data arising
from this model.

METHODS

Confluent monolayers of human endothelial cells
derived from umbilical vein or cardiac valves have been
grown and maintained in tissue culture as previously
described (3,4). The infection assay is performed as
described (3). Briefly, bacteria, adjusted to a fixed
density, are added to the endothelial monolayers and
incubated. Following incubation, the supernate is
decanted, the surface washed and the remaining adherent
bacteria removed with trypsin. Aliquots of this
suspension are serially diluted and plated into agar. The
number of adherent bacteria are determined by viability
counts of bacterial colonies.

Fc receptor expression was measured in infected and
uninfected endothelial cells using ^{51}Cr sensitized
sheep red blood cells coated with IgG as described (5).

RESULTS

Ultrastructural Study of S. aureus - Human Endothelial
Cell Interactions.

S. aureus adhere to human umbilical vein and cardiac
valve endothelial cells in higher numbers than other
bacterial species (3). When the assay is performed at 4°
C to prevent phagocytosis, binding occurs in a saturable
fashion in both dose and time response studies. Bacteria
are phagocytosed following adhesion. This process can be
blocked by pretreatment of the endothelial cells with
cytochalasin B and occurs with both viable and ultra-
violet killed staphylococci. Thus phagocytosis is an
endothelial cell mediated function. Quantitative studies
demonstrate that internalization of bacteria occurs at a
far slower rate than adhesion (6). Bacteria are
surrounded by pseudopods, enclosed in a membrane bound
vacuole and internalized (Figure 1).

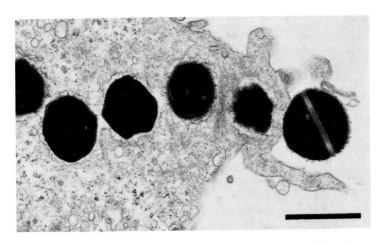

Figure 1. Internalization of staphylococci by human endothelial cells (1/2 hour incubation. Bar 1.0 um.) Reprinted with permission of Academic Press, Inc.

Once intracellular, bacteria appear to replicate despite fusion with lysosomes. Many of the steps shown with invasion of endothelial cells are also demonstrable with endovascular tissue obtained from rabbit cardiac valves or human aorta (6).

Endothelial Cell Modulation Prior to Infection.

We investigated whether modulation of human endothelial cells alters susceptibility to infection. Endothelial cell activation by interleukin 1 or endotoxin results in increased leukocyte adhesion (7). We studied the effects of pretreatment of the endothelial cell monolayer with acidic fibroblast growth factor (aFGF), interleukin 1 and endotoxin on subsequent S. aureus adhesion. aFGF decreased S. aureus adherence. This effect was maximally demonstrated following a 72 hour incubation. It was independent of the initial bacterial inoculum and was demonstrable with 3 of 4 S. aureus isolates tested (8). Interleukin and endotoxin were without effect following preincubation from 15 minutes to 48 hours (Table 1).

The effect of proteoglycans on S. aureus adhesion was assessed by treatment of the cells with the enzymes

TABLE 1
EFFECT OF ENDOTHELIAL CELL MODULATORS ON SUBSEQUENT
S. AUREUS ADHERENCE[a]

Modulator	Effect on S. aureus adherence
aFGF (purified human 50 ng/ml)	↓
Heparitinase (1.25u/ml)	↑
Endotoxin (1.0ug/ml)	↔
Interleukin 1 (10u/ml)	↔
Chondroitinase ABC (1.25u/ml)	↔

[a]Pretreatment of endothelial cell monolayers with the
modulator for varying time intervals was followed by
performance of the infection assay.

chondroitinase ABC or heparitinase (Table 1). Pretreatment of the endothelial cells with heparitinase caused a
significant increase in S. aureus adherence, suggesting
that heparan sulfate proteoglycans were involved in S.
aureus adhesion.

Endothelial Cell Modulation Following S. aureus Infection.

Fc receptor expression was significantly increased
following S. aureus infection of the endothelial cells.
The Fc receptor expression was inoculum dependent and was
blocked by aggregated IgG.

DISCUSSION

The versatility of S. aureus as a pathogen is demonstrated by its capacity to colonize and invade normal endovascular tissue. This enables staphylococci to cause life
threatening endovascular infections, often with metastatic
suppurative foci (1,2). The ability to maintain endothelial
cells in tissue culture has provided a method for studying
S. aureus - human endothelial cell interactions. We
have utilized an in vitro infection assay to explore
the steps involved in bacterial invasion. In our model
the sequence of events include adhesion, phagocytosis and
intracellular replication followed by progressive cellular
injury and eventually cell disruption.

Staphylococci adhere more avidly to the endothelial cell surface than other bacterial species. Cellular modulators including aFGF or heparan sulfate influence S. aureus adherence (Figure 2). aFGF has been shown to

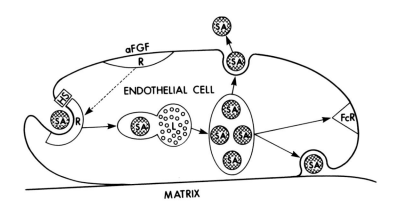

Figure 2. Model of S. aureus - human endothelial cell interactions, in vitro. HS, heparan sulfate; R, receptor; L, lysosomes; SA, S. aureus

play a critical role in endothelial cell function in vitro including cell growth, adhesion to substrate and glycosaminoglycan synthesis (9-10). Demonstration of its effect on staphylococcal invasion is a novel finding. Following bacterial infection, the endothelial cell is altered as shown by the expression of Fc receptors. Thus endothelial cell modulation influences infectability of the cell and infection of the cell also results in modulation of the cell.

These studies demonstrate the potential utility of this model to study this clinically important process. The model provides a unique method for detailed investigation of the sequential steps necessary for bacterial invasion. These studies have provided an alternative hypothesis for the pathogenesis of endovascular infections such as endocarditis. Staphylococci are capable of colonizing morphologically normal endovascular surfaces. Subtle biochemical modulations of these cells appear sufficient to alter the susceptibility of these cells to infection. This is in contrast to the case in subacute

endocarditis where a previously damaged valvular surface
is seeded by a relatively avirulent organism. Following
staphylococcal adhesion and phagocytosis, the bacteria are
provided with a relatively protected environment which
allows for replication and elaboration of toxin. Alterations of the endothelial cell may then contribute to the
progression of the disease. In endocarditis, this may
include vegetation formation, tissue destruction and
spread of infection to adjoining tissues.

REFERENCES

1. Waldvogel FA (1985). Staphylococcus aureus (including
    toxic shock syndrome). In Mandell GL, Douglas RG Jr,
    Bennett JE (eds) "Principles and Practice of Infectious
    Diseases, New York: John Wiley & Sons, p 1096.
2. Sheagren JN (1984). Staphylococcus aureus: the
    persistent pathogen. N Engl J Med 310:1368, 1437.
3. Ogawa SK, Yurberg ER, Hatcher VB, Levitt MA, and Lowy
    FD (1985). Bacterial adherence to human endothelial
    cells in vitro. Infect Immun 50:218.
4. Gordon P, Sussman II, and Hatcher V (1983). Long-term
    growth of human endothelial cells. In Vitro 19:661.
5. Cines DB, Lyss AP, Bina M, Corkey R, Kefalides NA, and
    Friedman HM (1982). Fc and C3 receptors induced by
    herpes simplex virus on cultured human endothelial
    cells. J Clin Invest 69:123.
6. Lowy FD, Fant J, Higgins LL, Ogawa SK, and Hatcher VB
    (1988). Staphylococcus aureus - human endothelial cell
    interactions. J Ultrastruc Mol Struc Res 98:In Press.
7. Bevilacqua MP, Pober JS, Wheeler ME, Cotran RS and
    Gimbrone MA (1985). Interleukin 1 acts on cultured
    human vascular endothelium to increase the adhesion
    of polymorphonuclear leukocytes, monocytes, and related
    cell lines. J Clin Invest 76:2003.
8. Blumberg EA, Hatcher VB, Lowy FD (1988). Acidic
    fibroblast growth factor modulates Staphylococcus
    aureus adherence to human endothelial cells. Infect
    Immun In Press.
9. Gospodarowicz D, Ferrara N, Schweigerer L, and Neufeld G
    (1987). Structural characterization and biological
    functions of fibroblast growth factor. Endoc Rev 8:95.
10. Gordon PB, Conn G, and Hatcher VB (1984). Glycosamino-
    glycan synthesis in cultures of early and late passage
    human endothelial cells: The influence of an anionic
    endothelial cell growth factor and extracellular matrix.
    J Cell Physiol 125:596.

Tissue Engineering, pages 115–120

PERSISTENT CLINICAL FUNCTION WITH PERCUTANEOUS DEVICES[1]

Lars M. Bjursten[2], Kajsa-Mia Holgers[2,3], Peter Thomsen[2],
Anders Tjellström[3], and Lars E. Ericson[2]

Department of Anatomy, University of Göteborg,
P.O.Box 33031, S-40033 Göteborg, Sweden

ABSTRACT A long term function has been established in clinical practice for transmucosal and percutaneous titanium implants. The present article focuses on important phenomena for the success of percutaneous devices based on information from clinical biopsies and experimental studies.

INTRODUCTION

Several problems are associated with the clinical use of implants penetrating the skin (percuatneous devices). These problems are manifested as infections and rejection phenomena, leading to implant failure (1). A procedure for installation of bone-anchored implants for attachement of fixed dental bridges has been clinically established by professor Brånemark and coworkers (2). The success rates in consecutive studies over 15 years exceed 95 percent for individual implants in the lower jaw (3).

The installation is based on a two-step surgical procedure, where the first step is a careful installation of a threaded fixture in the jaw-bone. The implantation site is meticulously prepared using low-speed drilling and threading of the bone, and the implant covered by the mucosa and

[1] This work was supported by grants from the Institute for Applied Biotechnology, The Swedish Board for Technincal Development, The Göteborg Medical Council and The Gustav V 80-year fund.
[2] Department of Anatomy, University of Göteborg
[3] Department of ENT, Sahlgrens Hospital, University of Göteborg

allowed to heal for 3 - 6 months. In a second surgical step
the fixture is exposed and a mucosa penetrating abutment is
screwed on top of the fixture. The artificial teeth are then
built on a bridge that is screwed on top of the abutments.

These principles have been adopted also for anchorage of
cranio-facial prostheses eg. silastic noses and ears (4). The
10 year clinical experience with these bone-anchored
percutaneous devices in the head and neck region are very
good with failure rates below 0.12 % per month (5). These
figures compare well with failure rates for eg. pacemakers but
are far better than for eg. peritoneal dialysis catheters
(6). The present study will discuss factors believed to be
important for the clinical success of these devices and
adress possible animal experimental models to study these
factors.

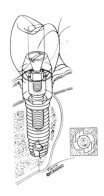

FIGURE 1. Schematic showing the assembled dental implant
consisting of the fixture anchored in the jaw bone, the
mucosa penetrating abutment and the artificial tooth.

RESULTS

Clinical Results

All 92 patients who have had percutaneous devices
installed at the ENT department, Sahlgrens hospital, Göteborg
between 1977 and may 1985 were included in the study. They
were regularly followed up by examinations every 3 months at
the Department. The degree of skin irritation was documented.
Slight irritation which was rapidly reversed by local

cleaning was noted at 5.1% of the appointments. A severe
irritation, needing local revison, was seen at 0.6% of the
observations. Problems with skin irritation did not increase
after long implant periods.

Clinical Biopsies

Biopsies were obtained from the skin adjacent to the
abutments for some patients with signs of inflammation but
also from clinically not irritated skin when the abutments
were removed for unrelated reasons. Biopsies from the same
area during installation of the abutments served as controls.
The biopsies were divided and processed for plastic embedding
and staining as well as cryostat sectioning. The specimens
showed (Figure 2) that a slight downgrowth of the epithelium
had occurred around the abutments. No direct attachement
between the epithelium and the abutment could be seen. Often
an intervening layer of filamentous material or inflammatory
cells were present. A contact could be observed between the
epidermal connective tissue and the abutment. Inflammatory
cells were present in clusters in the connective tissue at
some distance from the interface. When clinical signs of
inflammation was present an increased number of polymorphonuclear granulocytes were present at the implant surface.
    In the cryostat sections an increased number of B-
lymfocytes and antibody producing plasmacells could be
detected using immunohistochemical techniques and monoclonal
antibodies.

FIGURE 2. Tissue biopsy from clinically healthy skin
adjacent to implant, which has been on the right hand side.
Surface of the skin is facing top.

Experimental Percutaneous Implants

In order to study the mechanisms behind the healing of percutaneous implants and gain insight into which factors are important for the persistence of the clinically used percutaneous titanium devices an experimental model in the naked rat has recently been developed. The animal model comprises the installation of a percutaneous device in the parietal bone of the rat using a two step procedure. Preliminary results show that these devices can persist for at least half a year despite the absence of direct contact between implant and epithelium.

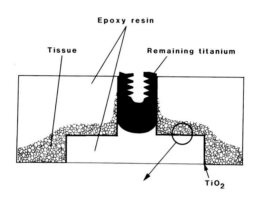

FIGURE 3. Schematic of electrochemical method to remove the titanium in implants embedded en bloc with tissue in order to make ultrathin sections of intact interface.

Tissue Response to Different Implanted Materials

Differences in tissue response to various implanted materials has been studied extensively over the years (see eg. 7). We have investigated the kinetics as well as the differences in interfacial tissue response to pure titanium and polymer implants inserted in the abdominal wall of rats (8,9). In order to study the intact tissue interface to titanium implants, a method was developed based on the electrochemical dissolution of the bulk titanium, leaving the titanium dioxide on the implant surface intact (Figure 3) (9). These studies reveal that the implant during the first day is surrounded by polymorphonuclear leukocytes which were later followed by macrophages and fibroblasts. This healing

period is characterized by an initial fluid space close to
the titanium surface. The space is gradually replaced by
organized tissue which approach the implant surface and the
implant becomes integrated (Figure 4). For the polymers so
far tested a different response was seen. Activated
macrophages remain at the implant surface, the fluid space
persist and a surrounding fibrous capsule is formed.

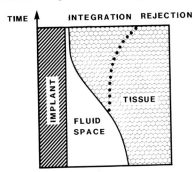

FIGURE 4. Schematic showing the kinetics of the tissue
organization at the implant interface leading to integration
or rejection. Redrawn after Kasemo and Lausma (10).

DISCUSSION

Clinical lasting function with tranmucosal and
percutaneous titanium implants has been achieved earlier
(3,5). The present study provides some information enabeling
a speculation on what mechanism and factors are important for
the tissue response to percutaneous implants.

Present results as well as previous findings (eg 7,8)
show that the selection of the implant material is highly
important and that pure titanium induces a favorable tissue
response – integration. The barrier function of the skin or
mucosa is impaired by the presence of the implant. The
presence of antibody producing cells and polymorphonuclear
leukocytes in the tissue around implants showing clinical
signs of inflammation indicate an increased exposure to
noxious substances and microorganisms. Clinical experience
shows that surgical reduction of subcutaneous tissue gives
less problems with skin inflammation (4). An experimental
model in the naked rat has been established in order to study
the kinetics of the tissue response to percutaneous implants.
This model also enabels the study of eg. the host defence and
response to implants and exogenous factors.

REFERENCES

1.   Winter GD (1974). Transcutaneous implants: Reactions of
      the skin - implant interface. J Biomed Mater Res
      Symposium No 5:99-113

2.   Brånemark P-I, Hansson B-O,, Adell R, Reine U, Lindström
      J, Hallén O, Öhman A (1977). Osseointegrated implants in
      the treatment of edentulous jaw. Scand J Plast Reconstr
      Surg 11 (suppl 16)

3.   Adell R, Lekholm U, Rockler B, Brånemark P-I (1981). A
      fifteen year study of osseointegrated implants in the
      treatmant of the edentulous jaw. Int J Oral Surg 10:387.

4.   Tjellström A (1985). Percutaneous implants in clinical
      practice. CRC Critical Reviews in Biocompatibility
      1:205-228.

5.   Holgers K-M, Tjellström A, Bjursten LM, Erlandsson BE
      (1987). Soft tissue reactions around percutaneous
      implants. Int J Oral Maxillofacial Implants 2:35-39.

6.   Kim D, Burke D, Izatt S, Mathews R, Wu G, Khanna R, Vas
      S, Oreopoulos DG (1984). Single or double-cuff
      peritoneal catheters? A prospective comparison. Trans Am
      Soc Artif Inter Organs 30:232-235.

7.   Williams DF (1982; ed): Biocompatibility of clinical
      implant materials I-II, Boca Raton, CRC Press.

8.   Thomsen P, Bjursten LM, Ericson LE (1986). Implants in
      the abdominal wall of the rat. Scand J Reconstr Surg
      20:173-182.

9.   Bjursten LM, Thomsen P, Eriksson AS, Olsson R, Ericson
      LE (1988) In Proc 7th European conference on
      biomaterials, Amsterdam, Elsevier.

10.   Kasemo B, Lausmaa J (1986). Surface science aspects on
      inorganic biomaterials. CRC Crit Rev Biocompat 2:335.

Tissue Engineering, pages 121–126

SOFT TISSUE ATTACHMENT AT AN IMPLANTED SURFACE[1]

C.A. Squier, A.W. Romanowski, and P. Collins

Departments of Oral Pathology, Periodontology and
Dows Institute for Dental Research
College of Dentistry
The University of Iowa, Iowa City, Iowa 52242

ABSTRACT   Prostheses   penetrating   skin   or   mucous
membranes involve junctions with both epithelium and
connective tissue. We have examined the role of
porosity of surfaces implanted in skin and determined
that surface pores greater than 3 μm in diameter
permit attachment of collagen fibers and minimize
epithelial downgrowth so that a stable junction is
formed. However, optimal attachment of stratified
squamous epithelium to a surface appears to require
the presence of a specialized junctional epithelium,
such as exists at the tooth-gingiva junction.
Evidence   suggests   that   the   formation   of   such
junctional epithelium depends on the presence of a
suitable attachment surface and the appropriate type
of   underlying   connective   tissue   to   promote
differentiation of this epithelium.

INTRODUCTION

In attempting to obtain an adequate interface between
body tissues and structures such as catheters or rigid
implants, many ingenious mechanical devices have been
proposed (1,2).   Consideration has also been given to
"natural" interfaces, such as those between skin and nail,
hair, or antler, or between the mouth lining and the tooth
(1,2,3). This latter junction, unlike the other examples,

[1]Supported in part by US PHS NIH 5 K16 DE00175

consists of an interface between structures with different mechanical and biological properties (tooth enamel and gingiva) and may provide a useful model for designing percutaneous and permucosal prostheses. The attachment to the tooth consists of two components; an epithelial attachment to the enamel surface and a connective tissue attachment in which collagen fibers are embedded in the calcified material of the root. The epithelial attachment is mediated by a unique "junctional" tissue, consisting of a poorly differentiated epithelium attached to the enamel surface via hemidesmosomes and a basal lamina (4). In defining the parameters necessary for a stable tooth-soft tissue junction we must consider the attachment of both the epithelial and connective tissue components at the interface as well as the interactions between them. A critical factor for cellular and tissue attachment is the nature of the attachment surface, including its porosity. We have examined this parameter in a model system involving the implantation of material of various porosity into the back skin of pigs. This approach avoids some of the unnecessary complications of working in the oral cavity, such as the extensive microbial flora and the mechanical trauma of mastication, yet provides a model of tissue behavior that may be extrapolated to other sites.

METHODS

The experimental procedure (5) involves implanting type FM (cellulose acetate) Millipore filters (Millipore Corp., Bedford, Mass.) with pore sizes ranging from 0.025 to 8 µm into the backskin of pigs. Full thickness biopsies of skin containing the filter are removed after intervals of 1, 2, 4, 6, or 8 weeks and processed for examination by light and transmission electron microscopy. Measurements are made of the extent of the epithelial downgrowth along the filter from the adjacent epidermal-connective tissue junction and the data plotted graphically to show the extent of downgrowth with time.

RESULTS AND DISCUSSION

The rate of epithelial migration along the filters was inversely related to pore size (Fig. 1). However, significant differences are evident between the filters with the largest pore sizes (3-8 µm) and the other groups. The downgrowth in the former group extended less than half the distance of that in the latter group, and after moving rapidly during the first week of the experiment formed a relatively stable epithelial-filter junction with little further migration for the rest of the experimental period.

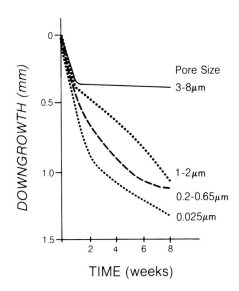

FIGURE 1. Rate of migration of epithelium along filter surfaces (based on data from Squier and Collins (5).

The difference in migratory behavior of the epithelium appeared to be associated with the ability of connective tissue cells to enter the interstices of the filter; no cells were present in the 0.025 µm pores, few were seen in the 0.65 or 1.0 µm pores and many were present within the 3, 7 and 8 µm pores. More significantly, there was evidence of collagen fibril formation within the interstices of these latter filters so that a network of collagen extended more or less at right angles to the filter surface in continuity with the fibers of the lamina propria (Fig. 2a. Ultrastructurally, these fibrils were intimately associated with cells within the filter interstices (presumably

fibroblasts) and, on occasion, also appeared to be attached
to the filter material (Fig. 2b). By contrast, collagen
around the filters with the smallest pore size never
appeared to penetrate the filter and was arranged in a
parallel pattern to form a fibrous capsule (Fig. 2c).

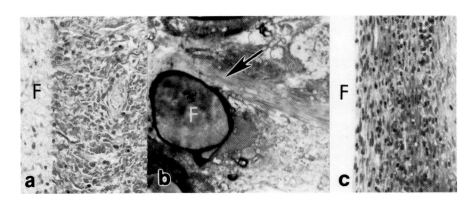

FIGURE 2. a) Light micrograph of collagen infiltrating
the interstices of a 8.0 μm filter; b) Electron micrograph
of 8.0 μm filter surface showing apparent attachment of
collagen fibrils to filter material (arrow); c) Light
micrograph of parallel arrangement of collagen at surface of
0.025 μm filter. F=filter; magnification a, c, x200 b,
x4000.

It seems clear that infiltration of the filter with
collagen effectively halted epithelial migration along the
filter surface. This has been demonstrated by Winter (3,6)
in a similar system and is in accord with observations at
the tooth-gingiva junction, where intact periodontal fibers
deter epithelial migration along the tooth root (7,8).
However, the existence of a stable epithelial junction does
not imply the existence of an epithelial seal, such as that
between the tooth enamel surface and the junctional
epithelium. Examination of the epithelium-filter interface
with the electron microscope showed that epithelial cells
migrated along the surface of the filter and did not enter
the interstices of even the coarsest filters, which were
filled with a dense amorphous coagulum. While forming a
stable junction adjacent to less porous filter surfaces, the

epithelium continued to show a maturation gradient from the basal lamina towards the filter surface such that a thin layer of keratinized cells approximated the filter (Fig. 3).

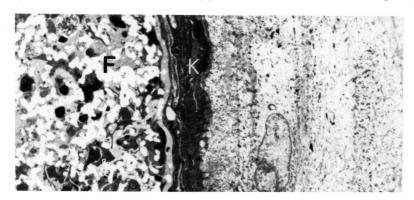

FIGURE 3. Electron micrograph of epithelial cells at surface of 0.65 μm filter--note presence of keratin (K) adjacent to filter (F). Magnification x8000

This is a very different arrangement from the poorly differentiated junctional epithelium at the dento-gingival interface that attaches to the enamel surface via hemidesmosomes and a basal lamina. It appears, therefore, that factors other than a connective tissue attachment and a relatively smooth interface surface are necessary to achieve a true junctional epithelium, and these may involve the molecular nature of the attachment surface and/or the existence of appropriate connective tissue influence.It has recently been suggested (9) that the maintenance of epithelial differentiation in the adult is a function of the underlying connective tissue and that the normal superficial connective tissue (lamina propria) permits the expression of a pattern of differentiation whereas deeper connective tissue does not have this ability and overlying epithelium fails to differentiate. The junctional epithelium overlays a "deep" connective tissue--the periodontal ligament--and accordingly shows little differentiation. Such assumptions may justify the recent attempts by periodontists to achieve the appropriate connective tissue adjacent to the tooth by guided tissue regeneration during healing after periodontal surgery (10).

In terms of implant surfaces these considerations mean
that a successful soft tissue interface will demand more
than selecting materials with appropriate surface properties
and biocompatibility and that attention will have to be
given to the connective tissue-epithelial interactions at
this site, possibly by the introduction of appropriate
connective tissue or cell types.

ACKNOWLEDGMENTS

We thank Ian Mackenzie for valuable discussion and John
Laffoon for assistance with the electron microscopy.

REFERENCES

1. Groose-Sietrup, C (1984). Design criteria for
   percutaneous devices. J. Biomed Mat Res 18:357.
2. von Recum, AF (1984). Applications and failure modes of
   percutaneous devices: a review. J Biomed Mat Res
   18:323.
3. Winter, GD (1974). Transcutaneous implants; reactions
   of the skin-implant interface. J Biomed Mat Res
   Symposium 5:98.
4. Schroeder, HE, Listgarten, MA (1977). "Fine structure of
   the Developing Epithelial Attachment of Human Teeth."
   (2nd ed.) S. Karger, Basel.
5. Squier, CA, Collins, P (1981). The relationship between
   soft tissue attachment, epithelial downgrowth, and
   surface porosity. J Periodont Res 16:434.
6. Winter, GD (1972). Epidermal regeneration studied in
   the domestic pig. In Mailbach HI and Rovee, DT (eds):
   "Epidermal Wound Healing," Chicago, Year Book, Medical
   Publications, Inc., pp 71.
7. Levine, L, Stahl, SS (1972). Repair following
   periodontal flap surgery with the retention of gingival
   fibers. J Periodontal 43:99.
8. Stahl, SS (1977). Healing following simulated fiber
   retention procedures in rats. J Periodontol 48:497.
9. Mackenzie, IC (1987). Nature and mechanisms of
   regeneration of the junctional epithelial phenotype. J
   Periodont Res 22:243.
10. Nyman, S, Lindhe, J, Karring, T, Rylander, H (1982).
    New attachment following treatment of human periodontal
    disease. J Clin Periodontol 9:290.

Tissue Engineering, pages 127–136
© 1988 Alan R. Liss, Inc.

Dystrophic Calcification and Mineralization During Bone
Induction: Biochemical Differences

Marcel E. Nimni

Laboratory of Connective Tissue Biochemistry,
Departments of Biochemistry and Medicine, University of
Southern California School of Medicine and Orthopaedic
Hospital, 2400 South Flower Street, Los Angeles,
California 90007

ABSTRACT The calcification of implants of
glutaraldehyde-crosslinked collagenous tissues and
collagen was studied in young and old rats and
compared to bone induction by non-crosslinked
osteogenically active demineralized bone matrix
(DBM). Glutaraldehyde-crosslinked implants of DBM,
tendon, and cartilage calcified in young but not in
old animals and accumulated only trace amounts of BGP
(Bone Gla protein, osteocalcin). Alkaline
phosphatase activity was high in implants of DBM and
undetectable in crosslinked implants. To try and
understand why bone formation is so significantly
reduced in older Fischer 344 rats, we developed a
system which consists of cylinders of DBM sealed at
the ends with a Millipore filter. Cells originating
from 20 day old embryo donors were introduced into
the chambers prior to subcutaneous implantation.
After 4 weeks of implantation in 26 month old rats,
the cylinders containing embryonic calvaria or muscle
cells were found to be full of bone and/or cartilage.

INTRODUCTION

Dystrophic Calcification

     Formation of bone at various stages of development is
an active process which involves the synthesis of

connective tissue macromolecules (collagen, bone specific proteins and proteoglycans) by osteoblasts, followed by deposition of a calcium phosphate mineral phase in close contact with the collagen fibrils. In contrast to this highly regulated process, mineral can deposit at ectopic sites and give rise to various pathological states. The sequence of events that accompany these two related processes and possible mechanisms involved are discussed in detail in Glimcher's review.

Of particular interest is the calcification of implants derived from chemically crosslinked collagenous tissues, such as glutaraldehyde-treated porcine heart valves or prosthestic valves made from bovine pericardium. We have shown that glutaraldehyde introduces thermal and chemically stable crosslinks into collagen fibers, while other aldehydes, such as formaldehyde, do not. This finding was extended to porcine heart valves to render them biologically stable and essentially non-antigenic, thus providing a suitable prosthetic replacement for diseased tissues.

The glutaraldehyde technology has been applied successfully to other bioprosthetic materials such as skin, tendons, ligaments, pericardial patches, and collagen implants. Several problems have arisen during the follow up of patients with collagen-derived implants, including fibrotic infiltration, matrix degeneration, calcification, immune rejection, sensitization, toxicity of unreacted glutaraldehyde and thrombosis.

Bone Induction in Older Animals

It is generally acknowledged that bone mass, bone remodeling and ability of fractures to heal are decreased with advancing age. Ectopic bone formation, induced by the implantation of demineralized bone matrix (DBM), also appears to be age dependent. Syftestad and Urist reported that acid demineralized bone matrix from older animals was less effective in generating an osteogenic response. Using a similar bone induction model, Irving et. al. showed that in 6-week old rats new bone forms 14 days post-implantation while deposition of new bone in 2-year old animals takes place 23 days after implantation.

We have recently confirmed and extended the above observations using Fischer-344 rats of various ages. Total calcium accumulation and the rate of ^{45}Ca

deposition was greatly depressed in the older animals.
The peak of alkaline phosphatase activity in older animals
was reached after longer periods of time and total enzyme
activity was lower in the 10- and 16-month old compared to
1-month old rats. Implants in older rats exhibited
reduced amounts of a bone specific vitamin K-dependent
Gla-protein, BGP. Significant differences in calcium
content were also observed between 1-month old and 3-month
old rats. These findings and the histological examination
of implants provided additional support for the decreased
ability of older animals to produce new bone in implants
of demineralized bone matrix.

In our attempts to enhance the bone forming potential
of older animals we tested the contribution to the bone
induction process of osteoprogenitor cells derived from
fetal tissues. In the present study we present our
findings on the enhanced osteogenic response of old rats
to the subcutaneous implantation of embryonic calvaria
cells enclosed in specially designed demineralized bone
matrix chambers.

MATERIALS AND METHODS

Details of experimental procedures for crosslinking,
assay of calcium, alkaline phosphatase and BGP
(osteocalcin), preparation of DBM chambers, isolation of
embryonic cells and preparation of demineralized bone
matrix (DBM) are provided in the references cited at the
end of the text.

RESULTS

Dystrophic Calcification

We have reported the tendency of crosslinked bovine
pericardium to calcify. We selected this tissue because
of its large proportion of type I and type III collagen
and because it is frequently used to construct heart valve
prostheses. As in all the implanted tissues tested,
calcification was much greater in young animals (1-month
old at implantation time) than in older ones (8-months
old). Due to the tendency of crosslinked pericardium to
calcify, even implants in older animals after a lag period
begin to accumulate large amounts of calcium which

approach levels comparable to those in young rats 16 weeks
post-implantation. This was not true for implants of
other tissues in this study. Calcium in young rats
reached maximum levels around 6 weeks post-implantation
and then began to decline. Based on histological
observations, the process of biodegradation of implanted
crosslinked pericardium appeared to be slower in the older
animals.

      Comparative data on the ability of decalcified rat
bone matrix (DBM) to induce the formation of new bone, and
its capacity to calcify after its bone inductive ability
is destroyed by glutaraldehyde fixation, are given in
figure 1. In this particular study only young animals
were used, since we have previously demonstrated that
aging dramatically reduces the rate of new bone formation.
As expected, new bone formation was associated with a high
calcium content and also with an accumulation of BGP. In
contrast, calcification around the inactivated
glutaraldehyde-crosslinked DBM was associated with only
trace amounts of BGP (Fig. 3).

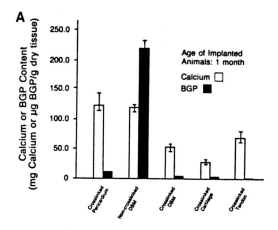

      Figure 1. Calcium and bone Gla protein (BGP)
contents in various glutaraldehyde-crosslinked collagenous
tissues compared to non-crosslinked demineralized bone
matrix (DBM) 4 weeks post-subcutaneous implantation in
1-month old rats. Error bars represent the standard error
of mean for seven separate implants.

Fig. 2 shows the relative abilities of various
crosslinked tissues to calcify when implanted in young
rats. At an early age the ability to calcify implants is
greatly enhanced compared to older animals.

Data on a classical parameter associated with bone
formation, i.e. alkaline phosphatase have also been
compared to calcium accumulation. Only new bone formation
induced by active DBM was accompanied by high enzymatic
activity, while dystrophic calcification as that in
implants of crosslinked pericardium was not (Fig. 3).

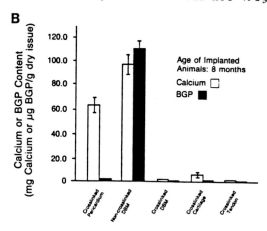

Figure 2. Same as figure 1 but in 8 month old rats.

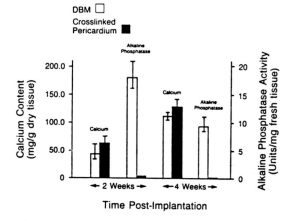

Figure 3. Calcium content and alkaline phosphatase
activity in implants of glutaraldehyde-crosslinked

pericardium compared to implants of inductively active,
non-glutaraldehyde-crosslinked demineralized bone matrix
(DBM) 2 and 4 weeks post-subcutaneous implantation in
2-month old rats. Error bars represent the standard error
of mean for seven separate implants.

Bone Formation in DBM Chambers

    Data on the osteogenic response to control cylinders
of demineralized bone matrix (DBM) and cylinders sealed at
both ends with Millipore filter, filled with calvaria
cells, after 4 weeks implantation in 26-month old animals,
are presented in Figure 4. While implants of DBM
cylinders (controls) in 26-month old rats accumulated only
minute amounts of calcium, bone Gla protein and alkaline
phosphatase, cylinders containing embryonic calvaria cells
exhibited high levels of all of these components contained
large amounts of histologically detected new bone (Fig.
5). Osteoclasts are very abundant in the actively
remodeling bone trabeculae. It is of interest that
calvaria cells in cylinders implanted in 26-month old
animals enhanced the amounts of new bone, assessed by
calcium content, to levels intermediate between those in
1-month and 8-month old rats, which had been implanted
with long segments of demineralized diaphysis.

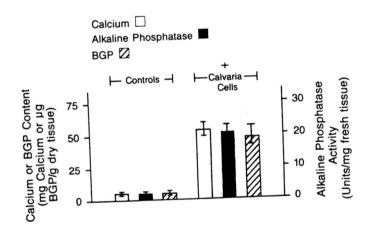

Figure 4. Calcium, BGP content and alkaline
phosphatase activity of implanted cylinders of

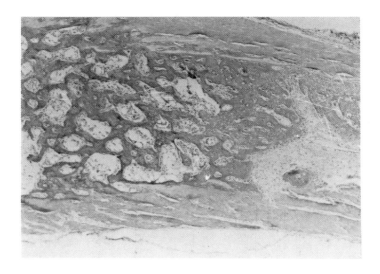

Figure 5. Section of an implant in which the marrow
space was filled with embryonic calvaria cells, and
removed 4 weeks post implantation, note that the spaces
are filled with newly formed trabecular bone x 50
        (Hematoxylin & Triosin).

We also studied the effects of transplanting various
sources of syngeneic embryonic cells. Fibroblasts were
ineffective, and possibly even inhibitory. Figure 6
summarizes a similar study using 9 month old rats
receiving calvaria cell transplants (10^6 cells),
calvaria cells devoid of periostial cells (10^5 cells)
and embryonic muscle cells obtained from the hind leg
(10^6 cells). All cell populations transplanted
enhanced significantly the levels of calcium ($p < .05$) and
alkaline phosphatase ($p < .01$) of the explanted chamber.

---

demineralized bone matrix (DBM) containing embryonic
calvaria cells, compared to controls filled with tissue
culture media, 4 weeks post-subcutaenous implantation in
26-month old Fisher rats.

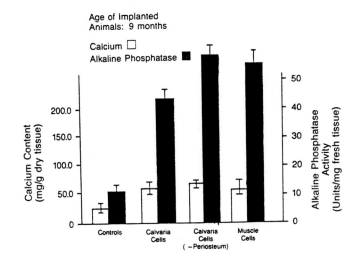

Figure 6.   Calcium contents and alkaline phosphatase activity within control DBM chambers containing either calvaria cells, calvaria cells devoid of periosteum or embryonic muscle cells. Rats were 9 month old at the time of implant and were explanted 28 days later.

DISCUSSION

Dystrophic Calcification

Bone formation during growth, remodeling or at ectopic sites under the stimulus of morphogenic factors is a cell mediated process. Dystrophic calcification, such as that in implants of bioprostheses containing crosslinked collagen (i.e. glutaraldehyde-treated porcine valves or bovine pericardial valves), involves deposition of calcium and phosphate by different mechanisms. Mineralization during bone formation is associated with high local levels of BGP and alkaline phosphatase which, however, is not the case for dystrophic calcification of crosslinked connective tissues. Calcification in bioprosthetic collagenous tissues is believed to be initiated by nucleation of hydroxyapatite within and outside collagen fibrils leading to deposition of mineral in the extracellular matrix of the bioprosthesis. It appears as if this process takes place in the absence of cells or matrix vesicles. The mineral deposited has been identified as crystalline hydroxyapatite similar to bone.

Age appears to play a major role in dystrophic calcification. Dystrophic calcification of crosslinked collagenous tissues is markedly age dependent, an observation made first with heart valve bioprostheses. This has raised concern about implanting such prostheses in young people. In our studies the material which showed the greatest tendency to calcify was crosslinked pericardium. The calcification of glutaraldehyde-treated aortic valves has been studied following subcutaneous implantation in rats, but such studies with crosslinked pericardial tissues have not, to our knowledge, been reported. Even older rats, which essentially fail to calcify glutaraldehyde-crosslinked cartilage, DBM and tendon, are able to calcify crosslinked pericardium to a significant extent. We are currently attempting to identify tissue components which may be responsible for the marked calcification of pericardium as opposed to tendon and other connective tissues.

Generation of Bone by Transplanted Embryonic Cells

Our results clearly demonstrate that cells released from 20 day old embryonic calvaria or muscle can generate bone when placed in DBM chambers and implanted subcutaneously in syngeneic senescent hosts. Control implants in such animals produce essentially no bone after 4 weeks. The degree of enhancement of bone production is very age dependant. Even though it is manifested in animals as young as 2 months of age it becomes significant in the mature and particularly in the senescent animals. Embryonic muscle cells were also capable of inducing bone formation, but embryonic skin fibroblasts were not. The latter tended to be inhibitory, probably by over growing or interfering with the osteogenic response induced by DBM.

The active chondrogenic and mild osteogenic response resulting from the implantation of embryonic muscle cells together with DBM were seen clearly. These events were inhibited when the DBM chambers were inactivated with trypsin prior to implantation.

We believe that our observations are novel and may provide a unique opportunity to further investigate the mechanism of bone induction and enhancement of new bone formation in old animals. Since DBM has less osteoinductive potential in older than in younger animals, it is likely that the effects observed in the present study are primarily associated with the osteoprogenitor or osteoinductive cells introduced. In our system bone marrow formed even in senescent rats, suggesting that the restraining chamber and/or the cell density achieved may be of fundamental significance. We are therefore further investigating these observations using other types of chambers, such as those made from enzyme-inactivated or guanidine-extracted DBM and various synthetic materials. We are also varying the modalities of application, such as suspensions in collagen or gelatin sponges or in agarose gels using embryonic muscle cells, bone marrow cells from young rats and calvaria cells from older animals.

REFERENCES

1. Nimni ME, Cheung D, Strates B, Kodama M, Sheikh K (1987). Chemically modified collagen: A natural biomaterial for tissue replacement. J Biomed Mat Res 21:741-771.
2. Nimni ME, Bernick S, Cheung DT, Ertl E, Nishimoto SK, Paule W, Salka C, Strates B (In press, 1988). Biochemical differences between dystrophic calcification of cross-linked collagen implants and mineralization during bone induction. Calcif Tissue Int.
3. Nimni ME, Bernick S, Ertl D, Nishimoto SK, Paule W, Strates BS, Villanueva J (In press, 1988). Ectopic bone formation is enhanced in senescent animals implanted with embryonic cells. Calcif Tissue Int.
4. Nishimoto SK, Chang C, Gendler E, Stryker WF, Nimni ME (1985). The effect of aging on bone formation in rats: Biochemical and histological evidence for decreased bone formation capacity. Calcif Tissue Int 37:617-624.
5. Bernick S, Paule W, Ertl D, Nishimoto SK, Nimni ME (In press, 1988). Cellular events associated with the induction of bone by demineralized bone matrix. J Orthopaedic Res.

Tissue Engineering, pages 137–143
© 1988 Alan R. Liss, Inc.

CHEMICALLY MODIFIED COLLAGEN: A NATURAL BIOMATERIAL FOR
TISSUE REPLACEMENT

Marcel E. Nimni and David T. Cheung

Laboratory of Connective Tissue Biochemistry,
Departments of Biochemistry and Medicine, University of
Southern California, School of Medicine, and Orthopaedic
Hospital of Los Angeles, Los Angeles, California 90007

ABSTRACT Glutaraldehyde crosslinking of native or
reconstituted collagen fibrils and tissues rich in
collagen significantly reduces biodegradation. Other
aldehydes are less efficient than glutaraldehyde in
generating chemically, biologically, and thermally
stable crosslinks. Tissues crosslinked with
glutaraldehyde retain many of the viscoelastic
properties of the native collagen fibrillar network
which render them suitable for bioprostheses.
Implants of collagenous materials crosslinked with
glutaraldehyde are subject long-term to
calcification, biodegradation, and low-grade immune
reactions. We have attempted to overcome these
problems by enhancing crosslinking through (a)
bridging of activated carboxyl groups with diamines
and (b) using glutaraldehyde to crosslink the
ε-NH$_2$ groups in collagen and the unreacted
amines introduced by aliphatic diamines. This
crosslinking reduces tissue degradation and nearly
eliminates humoral antibody induction. Covalent
binding of diphosphonates, specifically
3-amino-1-hydroxypropane-1, 1-diphosphonic acid
(3-APD), and chondroitin sulfate to collagen or to
the crosslink-enhanced collagen network reduces its
potential for calcification. Platelet aggregation is
also reduced by glutaraldehyde crosslinking and
nearly eliminated by the covalent binding of
chondroitin sulfate to collagen. The cytotoxicity of
residual glutaraldehyde leaching through the
interstices of the collagen fibrils or the tissue

matrix and of reactive aldehydes associated with the
bound polymeric glutaraldehyde can be minimized by
chemical neutralization and thorough rinsing after
crosslinking and storage in a nontoxic bacteriostatic
solution.

INTRODUCTION

Collagen is the most abundant mammalian protein,
accounting for around 30% of all body proteins. It is
deposited rapidly during periods of accelerated growth and
its rate of synthesis declines with age and in tissues
which undergo little remodeling. The intracellular
synthetic process, which leads to the formation of
procollagen, is relatively complex involving a series of
post-translational modifications. The catabolic pathway
is modulated by the activity of the enzyme collagenase
(1).
      The process of maturation by crosslinking imparts to
the collagen fibrils mechanical and biological stability.
Collagen can be extracted from tissues and reconstituted
into fibrils. These fibrils as well as tissues such as
tendons, heart valves, or pericardium, can be stabilized
by crosslinking and used for bioprostheses. This concept
prompted us to attempt to use collagenous tissues and
reconstituted and chemically modified collagens as
bioprosthetic materials (2).
      Our interest in bioprostheses began in 1966
subsequent to the observation that glutaraldehyde
introduces stable crosslinks into collagen fibers. Other
aldehydes were not as effective as glutaraldehyde in
generating thermally and chemically stable crosslinks. In
1969, it was suggested by Hancock (personal communication)
that the porcine heart valve, if properly processed, could
be a suitable prosthetic replacement for a failing human
valve. In view of earlier unsuccessful efforts to use
formaldehyde as a crosslinking agent, various agents were
tested for their ability to stabilize the collagen
framework. Porcine aortic valves with surrounding muscle
and connective tissue were treated with several aldehydes,
including formaldehyde, acetaldehyde, and glutaraldehyde,
under various conditions of temperature, ionic strength
and pH. Upon continuous exposure to running water at room
temperature treated tissues showed macroscopic and
microscopic structural degeneration except for those

treated with glutaraldehyde at neutral pH. Subsequent
biomechanical tests, heat of denaturation with associated
shrinkage and enzymatic degradation studies indicated that
glutaraldehyde-crosslinked bioprostheses were suitable for
in vivo studies and lead to the implantation of the
first such valve later that year in humans. This approach
has since been successfully applied to other bioprosthetic
materials, such as skin, tendons and ligaments,
pericardial patches, blood vessels and reconstituted
collagen.

Studies on the chemistry and reactivity of
glutaraldehyde with model compounds and monomeric and
polymeric collagens were followed by attempts to
understand the biological compatibility of
glutaraldehyde-crosslinked collagen. Other attempts were
also made to chemically modify the surface of the fibrils
and covalently attach macromolecules to the insoluble
matrix. Moreover, a serum protein (albumin), a
glycosaminoglycan (chondroitin sulfate) and a
diphosphonate (3-amino-1-hydroxypropane-1, 1-diphosphonic
acid or 3-APD) were tested for their ability to inhibit
calcification by in vitro nucleation and calcium
uptake tests and implantation in animals.
Biocompatibility tests were also conducted in tissue
culture and implants in animals of different ages. The
blood-bioprostheses interface was studied by scanning
electron microscopy and platelet aggregation tests. The
biological stability of glutaraldehyde-crosslinked tissues
and reconstituted and chemically modified collagens was
investigated histologically and by enzyme digestion
studies. Finally, crosslinked bovine collagen was tested
for immunogenicity in rabbits using procedures
specifically developed for detecting a highly insoluble
crosslinked antigen.

The studies summarized here are aimed towards further
understanding the chemistry and biology of collagen
in relation to the properties of bioprostheses.

RESULTS AND DISCUSSION

Mechanism of Collagen Crosslinking by Glutaraldehyde

Glutaraldehyde reacts primarily with amino groups of
proteins. Based on the spectral characteristics and the

molecular weights of the reaction products, it was predicted that glutaraldehyde will react to form an intermediate amine with a molecular weight of about 200 Daltons and absorption at 300 nm. In the presence of excess glutaraldehyde, the intermediate is quickly converted to a larger molecular weight intermediate which absorbs strongly at 265 nm. Larger intermediates were then altered to yield a strong absorption peak at 325 nm with no apparent change in molecular weight. These observations suggested that polymerization is induced by the initial reaction of glutaraldehyde with amines. The glutaraldehyde polymer-amine complex is selflimiting in size and can undergo internal rearrangement to become chemically inert.

Crosslinking of Collagen in Tissue Matrices

In the fixation of tissues or of densely packed molecules, as in reconstituted collagen fibers, a special concern has been the penetration of the glutaraldehyde molecules and accessibility of reactive protein groups. Incomplete crosslinking of the collagenous network can lead to enhanced biodegradation, antigenicity and loss of mechanical function.

Chemical Modification of Tissue Collagen and Reconstituted Collagen Fibrils to Enhance Biocompatibiltiy

Various animal tissues with ideal configuration are available for use as allogeneic or xenogeneic prostheses. Tissues, such as heart valves, fixed with glutaraldehyde can remain functional for many years. Failure eventually ensues because of deterioration of the collagenous-proteoglycan matrix resulting presumably from mechanical wear and tear or a foreign body reaction. The immune response may manifest itself as a low-grade immunological reaction with matrix degradation, exposure or generation of thrombogenic surfaces and calcification. In the latter case, crystal growth can contribute to hardening and enhanced tissue tear.

The types of crosslinks and modes of attachment of various crosslinking agents to tissues or reconstituted collagen networks are presented schematically in Figure 1.

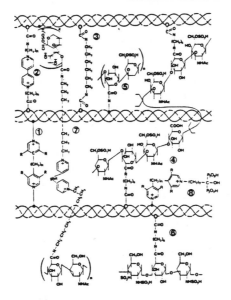

Figure 1.

● LYSINE
○ ASPARTIC or GLUTAMIC ACIDS

Glutaraldehyde crosslinks lysine residues of collagen and
the protein backbone of proteoglycans (Fig. 1, No. 1).
The R groups are OCH-(CH$_2$)$_2$ residues. Available
evidence suggested that a dipyridine structure may be
involved. The crosslink is of a "minimum size" compared
to the large glutaraldehyde polymers which may increase
the molecular size and length of the collagenous
structure.

In an effort to enhance the number and types of
crosslinks, the carboxyl groups of aspartic and glutamic
acids in peptides were modified and elongated as new
anchorage sites for glutaraldehyde crosslinking (Fig. 1,
No 2). New amino groups were introduced by the attachment
of an aliphatic diamine to the carboxyl groups via
carbodiimide activation. Glutaraldehyde is then attached
to these new amino groups situated more remotely from the
fibrous backbone. During the carbodimide activation
reaction crosslinking can also occur between two carboxyl
groups of neighboring collagen molecules via a diamine
bridge (Fig. 1, No. 3).

Anionic polysaccharides are attached via activated
carboxyl groups to residual peptide-bound lysines
unmodified by glutaraldehyde or to amino groups introduced
by diamine following reaction with peptide-bound carboxyl

groups (Fig.1, No.4). These groups can attach to various
sites on the chain or remain anchored by one end, as in
site No.5, Figure 1. A single attachment of the
carbohydrate moiety to the collagenous backbone may not
impart mechanical stability via crosslink formation. It
could, however, serve other functions, such as inhibition
of calcification, fill voids between the collagen fibers
and provide resiliency to the tissue. In general, growth
of hydroxyapatite crystals seems to be favored by the
presence of large water spaces between fibers and
inhibited by anionic mucopolysaccharides.

The diphosphonate 3-APD, which contains a reactive
amino group, may also be attached to the ε-NH$_2$ group
of lysine via glutaraldehyde (Fig. 1, No. 8). This
attachment has been shown by us and others to act as an
inhibitor of calcification in glutaraldehyde stabilized
collagenous matrices.

Heparin could be covalently attached to the
polypeptide backbone via carbodiimide activation of
carboxyl groups. In addition to this form of attachment,
heparin can be attached covalently to collagen by
glutaraldehyde. These two modes of attachment can
contribute to the maintenance of the antithrombogenic
properties of heparin and render it particularly useful as
a surface coating for bioprostheses. Heparin may be
crosslinked by the mechanism indicated in No.6, Figure 1.
Moreover, a collagen-proteoglycan complex, which includes
segments of diamine, glutaraldehyde polymers and anionic
polysaccharide chains, may be introduced to fill spaces
and displace unstructured water thereby interfering with
mineral nucleation and crystal growth while enhancing the
viscoelastic properties of the matrix (Fig. 1, No.7).

The diphosphonate 3-APD, which contains a reactive
amino group, may also be attached to the ε-NH$_2$ group
of lysine via glutaraldehyde (Fig. 1, No. 8). This
attachment has been shown by us and others to act as an
inhibitor of calcification in glutaraldehyde stabilized
collagenous matrices.

CONCLUSIONS

The stabilization of collagen fibers in xenograft
bioprostheses by crosslinking has been associated with the
development of an extremely valuable technology. The
experimental crosslinking approach has been extended to
purified collagens which can be reconstituted into fibrils
with desirable properties and made available in a variety
of shapes for use as bioprosthetic materials.

Current knowledge of the chemical and biological
properties of collagen can be used with advantage in the
preparation of suitable bioprostheses. For example, the

hemostatic properties of collagen can be utilized to
prepare microcrystalline collagen which can be sprinkled
on internal organs after surgery to stop bleeding.
Conversely, these properties may present a major handicap,
as in the case of clot formation associated with collagen
vascular prostheses. Proper understanding of the
mechanisms involved will allow us to modify the collagen
matrix and enhance biological compatibility.

Knowledge on how collagen nucleates hydroxyapatite
crystals from ionic calcium and phosphate solutions can be
utilized in bone induction experiments and in devising
ways to inhibit calcification in collagen-derived
bioprostheses. Control of biodegradation is of paramount
importance especially when the host uses the implanted
bioprosthesis as a scaffold for new tissue formation.
Tissue inertness is also a highly desirable feature
especially when the bioprosthesis is used for replacement
as in the case of artificial tendons and ligaments.

Significant advances have been made in understanding
the mechanism of crosslinking by the introduction of both
native and synthetic crosslinks using bifunctional agents.
Tissue immunogenicity can be reduced by modification of
the macromolecules using blocking agents and enzymatic
methods to remove antigenic determinants and rendering
them insoluble or unrecognizable by the immune system via
crosslinking. Other important properties, such as the
potential of collagen for cell attachment and interaction
with other macromolecules such as proteoglycans and
noncollagenous proteins, constitute additional areas of
considerable present and future interest.

REFERENCES

1. Nimni ME (1983). Collagen: Structure, function, and
   metabolism in normal and fibrotic tissues. Semin
   Arthritis Rheum 13:1-86.
2. Nimni ME (1988). "Collagen: Biochemistry,
   Biotechnology and Molecular Biology" Vol. I, II & III.
   Boca Raton, Florida: CRC Press.

Tissue Engineering, pages 145–150
© 1988 Alan R. Liss, Inc.

METABOLIC MODULATION of CELL INTERACTION in BLOOD, SPHEROIDS and TISSUES

Haim I. Bicher, M.D.

Director

Valley Cancer Institute, 14427 Chase St., Panorama City,
Los Angeles, CA 91402

The interaction of cells in a liquid milieu or in organized tissues is exquisitely dependent on the microenvironmental conditions of nutrient supply, metabolic removal and hormonal, autacoid and growth factor regulation. We have included all these factors under the term "metabolic regulation" and defined cell aggregation and growth behavior in several subsystems, as follows:

  A)   BLOOD CELL AGGREGATION: Both platelets and red cells have the ability to form clumps under normal or pathological conditions as part of the physiological process of hemostasis. When this reaction is exaggerated, then a tendency for thrombosis occurs, now established as an important part of the genesis of myocardial infarction and similar conditions. For the sake of this discussion we will consider the adhesive mechanism of each cell type separately, as they provide interesting models for active and passive aggregation.

  1.   Red Cell Aggregation: "Sludge" is the intravascular aggregation of red cells into masses, slowing the peripheral circulation and plugging up capillaries.

        The mechanism of erythrocyte aggregation involves the interaction of blood proteins with red cell membranes. The "suspension stability" of blood in vitro, as commonly determined when using the erythrocyte sedimentation rate in clinical practice, is therefore related to plasmatic factors, red cell factors, and the relationship between them.

        As plasmatic factors, the proteins play an all-important role. Apparently there is a correlation between the molecular weight of proteins and their ability to cause aggregation; the heavier proteins, such as macroglobulins and fibrinogen, being the strongest aggregation inducers. On the other hand, the albumins, of much smaller molecular weight, tend to inhibit aggregation.

        The number, morphology, and hemoglobin content of the red

cells can affect their aggregation tendencies, and consequently the rate at which they will sediment.   When hematocrit values diminish, sedimentation is faster.   Malformations prevent sedimentation.   The surface charge of the cells is also involved. Loss of sialic acid from the coating proteins diminishes the erythrocyte surface charges and increases aggregation tendencies. In vivo red cell aggregation seems to depend on a number of biochemical and physical factors. Among the biochemical factors, it is known that plasma proteins change in the presence of a massive sludge, and some of them actually coat the red cells, holding them together. The nature of this coating material is unclear, but globulin fractions and fibrinogen seem to be part of it.   Diseases in which plasma proteins are abnormal, such as macroglobulinemia, are characterized by severe blood   sludging. Increased plasma levels of fats and cholesterol also increase intravascular aggregation.

The physical appearance of circulating red cell masses   gives some clues as to the mechanisms involved.   Basically it can be distinquished between small, loosely held "rouleaux" type aggregates, and the more massive, unbreakable, deoxygenated clumps of a severe sludge.

Two physical factors widely investigated in relation to erythrocyte aggregation are the net electrical surface charge and blood viscosity.   As a general statement it can be said that during aggregation the normal, slightly negative charge of the erythrocyte is diminished, and in some cases reversed. Blood viscosity is usually increased, this being one of the ways in which sludges alter the normal circulation.

2.   Platelet Aggregation: Platelets play a central role in the mechanism of thrombosis, depending not only on their availability, but also on their capacity to adhere to each other or to damaged endothelium (2). Increased platelet adhesiveness has been described as an almost constant sequence of cellular destruction anywhere in the body, and considered to be of etiological significance in most thrombotic processes.   In contradistinction, decreased aggregation responses have been proven in several hemorrhagic diseases, such as hemorrhagic thrombocythemia, uremia and thrombasthenia.   The platelet defect in von Willebrandt's disease is still controversial.

Platelets aggregate under a variety of conditions, making it difficult to determine the physiological factor or combination of factors that will ultimately clump the platelets in a patient suffering from a thrombotic disorder.

In vitro, adenosine diphosphate (ADP) has been given the largest attention as an aggregation inducer.   Other substances such as thrombin, adrenaline, collagen, serotonin, bacterial endotoxin, the enzyme neuraminidase, phospholipids, free fatty acid and ionic calcium were reported among others as factors producing thrombocyte aggregation.

Most of these different substances seem to act through release of the endogenous ADP contained in the platelet.   Several investigators have forwarded this theory in respect to the clumping effect of thrombin, collagen, serotonin and adrenaline. They interact with the platelet membrane in such a way as to initiate the "release response," an active phenomenon by which several substances normally contained in the platelet, such as ADP, serotonin, lysozymes, and ionic potassium, are liberated into the surrounding medium, thus activating further the chain of events leading to hemostasis or thrombus formation. Even ADP itself can trigger the "release response," thus producing a two wave aggregation, the first one inititated directly by the ADP applied to the solution, and the second one induced by the release of endogenous ADP. Similar two wave responses are produced by adrenaline and thrombin. Substances that act only by liberating ADP from the platelet produce a delayed, one wave aggregation. This is the typical effect of collagen.

All the above mentioned responses have been induced using platelet rich plasma tested in the light aggregometer. When using the membrane capacitance aggregometer in whole blood, it can be detected that the effect of collagen on the platelet membrane is immediate. The platelet-to-collagen reaction is thought to represent the physiological mechanism by which platelets adhere to damaged endothelium, thus accomplishing their main function, that of hemostasis (2).

5-HT seems to be stored in the platelet in osmophilic granules. Some kind of specific 5-HT receptor has been forecast, since anti-serotonin drugs inhibit 5-HT-induced clumping, but not ADP-induced clumping. The specific role of serotonin in aggregation is unclear, but it seems to potentiate the effect of ADP. Catecholamines, especially epinephrine and norepinephrine also potentiate ADP clumping.

Several studies have demonstrated that a plasma protein, missing in von Willebrandt's disease and overactive in the plasma of diabetic patients, is necessary as a cofactor in ADP-induced aggregation. The exact nature of this protein is still controversial; it is different from fibrinogen, although a role for fibrinogen and its split products has also been suggested. Another protein clearly involved in the aggregation process and now under study is thrombosthenin, the contractile protein present alongside the platelet microtubules, and also associated with the platelet membrane, where it is almost indistinguishable for Ca++ activated ATPase. Current concepts of the mechanism of aggregation involve a complicated process of interaction between plasma and platelet proteins, requiring the presence of Ca++, Mg++ and perhaps other ions, and activated by the presence of energy-rich phosphates and certain enzymes. Of the energy-rich phosphates, ADP induces aggregation, while AMP, ATP and some other derivatives seem to inhibit it.

In vivo, hyperlipemia or increased levels of fatty acids of β-lipoproteins increase adhesiveness. During aggregation there is an increase in energy consumption, supplied at least in part through increased glycolysis.

The platelet has also a property to engulf and destroy foreign substances, and an active mechanism for lysosome activation. It has been suggested that activation of lysosome enzymes during or following aggregation plays a role in initiating viscous metamorphosis, thus connecting aggregation with the ensuing reactions of the clotting mechanism.

Finally, several physical factors are involved in the aggregation process. The shape of the platelets changes from that of a disc to that of a sphere with irregular surfaces and a variety of protruding processes. This is also associated with an increase in volume and a decrease in the normal negative surface charge (1,2).

B)   SPHEROIDS: Spheroids are in vitro growths of tumor cells in tissue culture that clump together to form a model spherical mass that develops concentrically.   After they attain a certain size (over 350µ diameter) they usually develop a necrotic center, while the peripheral regions remain well oxygenated (3,4,7).

In our studies we used spheroids of EMT6/Ro-cells.   Spheroid growth is initiated from single cells in the exponential growth phase by incubating suspensions of these cells in microbiological petri dishes for 4 days.   Spheroids are then transferred into spinner flasks where they are kept in stirred medium gassed with 3% (v/v) CO_2 and air.   Eagle's basal medium (BME) supported with 10% (v/v) newborn calf serum (NCS) serve as culture medium.   Spheroids with diameters in the range of 357µ to 2109µ entered our investigation.   Further details concerning the spheroid culturing are described elsewhere (3).

MEASURING SYSTEM: Multicellular tumor spheroids are assayed for their oxygenation status using O_2-sensitive microelectrodes in an experimental set-up that has been described in detail elsewhere (3,8).   Oxygen tension (pO_2) values are recorded in the spheroids on radial tracks under conditions that are similar to the growth conditions for these tumor cell aggregates in the spinner flasks.   Similarly, pH and glucose microelectrodes have been used to determine the influence of cell microenvironment on tissue growth and the formation of the necrotic center. It has been determined that the O_2 distribution in spheroids is graded towards the center, with areas of higher oxygenation in the periphery. The cellular growth fraction corresponds to the pO_2 concentration, being higher in the better oxygenated areas.   The effect of pH is correspondingly reversed, with lower pH in the necrotic region.

Similar studies have been done correlating spheroid growth in media of different glucose concentration (5).   Spheroids grown under higher glucose levels had increased viable volume but a higher percentage of quiescent cells with probable better diffusion inside the spheroid.   The necrotic center was reduced in the high glucose grown models, with correspondingly smaller radiation hypoxic fractions and increased radiation sensitivity, indicating better oxygenation.

It has also been shown that increased concentrations of amino acids and vitamins in the DMEM medium caused similar improvements in spheroid growth patterns.   However, the effect of high glucose concentration is less obvious on the less enriched media.   This is obviously a viable model to experiment on the effect of growth factors on tissue formation, as well as determining the needed levels of life supporting metabolites (3,5,6).

C) CONTROL OF LOCAL OXYGEN REGULATION MECHANISMS IN TISSUES: Brain tissue was chosen as a model of local metabolic control, since the tissues of the central nervous system are acutely sensitive to the effects of circulatory insufficiency. Interruption of the cerebral circulation is followed within seconds by a loss of consciousness and within minutes by irreversible changes in the brain.

It is because the maintenance of normal cerebral function is so vitally essential and completely dependent on an adequate blood supply that the study of cerebrovascular regulatory mechanisms and the effect of drugs upon these assume a special significance.

In several publications (8,9,10,11) we have described a precisely controlled autoregulatory mechanism to maintain a constant brain TpO_2 (tissue pressure of oxygen, which refers to the partial pressure [in mm Hg] of this gas at the measuring tip of the electrode) level. This includes a theoretical "tissue oxygen sensor" able to simultaneously regulate cerebral blood flow (CBF) and tissue O_2 consumption through reflex inhibition-excitation of generalized neuronal activity; it is defined by four criteria in the physiological compensatory response elicited by a short period of anoxic anoxia: 1. short "reoxygenation time" (RT, the time required for TpO_2 to return to the pre-anoxic level), 2. increase in CBF, 3. presence of an "overshoot" (period after reoxygenation during which TpO_2 is higher than baseline), and 4. a period of "anoxic silence" in neuronal activity.

All four responses can be suppressed by the α-adrenergic blocking agent, phenoxybenzamine. In the more recent experiments other autonomically active agents were tested, and a fifth autoregulation criterion, the small or non-existing rise in brain TpO_2 during normoatmospheric O_2 breathing, was added as another test. Also, the action of high respiratory CO_2 levels with these mechanisms was demonstrated. Atropine, propranolol and isoproterenol did not influence this specific mechanism (9).

Other organs, such as liver and kidney, are also able to control local tissue oxygen levels. These mechanisms are crucial to organ viability and can represent both models for successful new tissue microcirculatory organization or tests to ascertain survival characteristics under environmental or other stress.

D) A GLIMPSE OF THE FUTURE: Knowledge of metabolic controlling factors in cell adhesion is central to start growth of tissue in vitro or in matrix arrangements. Should these be programmable, the initiation and development of complex tissues will be done with clear target functions in mind, optimizing from the onset of the cellular culture stage the functions eventually required of the growing organ or prosthesis.

The spheroid model is particularly suitable for studying the relations between metabolic control and tissue growth. It will in all likelihood develop into the initial step for organ formation from single cell banks. This concept if fully applied could lead to the development of a process in which spheroidal clumps of different cells will be added to each other under varying controlling metabolic conditions to create multi-function organoids.

The survival of these organoids under IN VIVO stress conditions will there be tested by their metabolic autoregulation properties after biogenic microcirculation systems have been induced. Medical applications for these physiologically adaptable organoids could range from functioning pancreas, liver, kidneys to specific function brain nuclei or nerves.

REFERENCES

1. Bicher HI (1972). Blood cell aggregation in thrombic processes. Springfield: Ed. W Seegers, Charles C Thomas.
2. Bicher HI (1970). anti adhesive drugs in thrombosis. Thromb et Diath Haem. 42:197-214
3. Kaufman N, Bicher HI, Hetzel FW, Brown M (1981). A system for determining the pharmacology of indirect radiation sensitizer drugs on multicellular spheroids. Cancer Clinical Trials IV:199-204.
4. Busch NA, Bruley DF, Bicher HI (1982). Identification of viable regions in "in vivo" spheroidal tumors: a mathematical investigation. Adv Exp Med Biol 157:1-7.
5. Luk CK, Sutherland RM (1987). Nutrient modification of proliferation and radiation response in EMT6/Ro spheroids. Int J Rad Onc Bio Phys 13:885-895.
6. Wolfgang MK, Vaupel P (1987). Improvement of tumor spheroid oxygenation by tetrachlorodecaoxide. Int J Rad Onc Bio Phys 13:49-54.
7. Kaufman N, Hetzel FW, Bicher HI (1980). The effect of hyperthermia on oxygen distribution in spheroids. Radiat Res 83:395.
8. Bicher HI (1977). Modern methodology in the study of microphysiologic functions. Exerta Medica International Congress Series No. 399 Anesthesiology (ISBN 90 219 0327 0:488-492).
9. Bicher HI, Marvin P (1976). Pharmacological control of local oxygen regulation mechanisms in brain tissue. Stroke 7:469-472.
10. Bicher HI, Reneau DD, Bruley DF, et al (1973). Brain oxygen supply and neuronal activity under normal and hyperglycemic conditions. Am J Physiol. 224:275-282.
11. Bicher HI, Bruley DF, Reneau DD, et al (1971). Effect of microcirculation changes on brain tissue oxygenation. J Physiol 217:689-707.

Tissue Engineering, pages 151–152
© 1988 Alan R. Liss, Inc.

Summary/Discussion

Biomaterial Centered Infections, Implants and Inflammation Reactions

Randall Swartz, C. Fred Fox and Christopher A. Squier

Control of bacterial infection is a parameter that must be weighed when considering factors that maximize tissue compatibility of prostheses. Major problems of biomaterial- centered infections derive from both device complexity and asepsis. The selection of biomaterials for use in complex implants such as the artificial heart, and practices utilized in and subsequent manufacturing and handling must assure maintenance of long-term asepsis.

Anthony G. Gristina noted the problem which arises when microbes growing on surfaces of implants attach and secrete protective slime layers. The pockets of infection created are particularly difficult to treat because antibiotics encounter a diffusional barrier and cannot reach the bacteria in sufficient concentrations to be effective. These pockets serve as a chronic source of reinfection. These problems are reduced as compatibility between implant and tissue increases. If the biomaterial can form a tight junction with normal tissues, the bacteria can be excluded.

Discussion centered around improved surface properties, surface modifications for improved tissue compatability, and incroporation of antibiotics on or in prosthetic surfaces. Improved tissue compatability would prevent bacteria from attaching and establishing.

The in vitro model of Staphylococcus aureus adhesion, endocytosis and replication in human endothelial cells has significant clinical implications and was discussed by V.B. Hatcher. The capacity of S. Aureus to colonize and invade cultured "normal" endothelial cells in heart valves poses several interesting questions. What is the effect of the mechanical environment, including flow, on S. aureus to colonize and invade cultured human endothelial cells? Does the mechanical environment modulate the phenotypic expression of the endothelial cells so that there is increased S. aureus adhesion to the endothelial cells? In the future, the role of the mechanical environment on S. aureus adhesion to human endothelial cells should be investigated.

Lars M. Bjursten and Christopher Squier discussed the interaction of implants with surrounding tissues and the critical importance of surface interactions. Titanium implants gave outstanding performance with percutaneous applications; a normal fibronectin fibrin-collagen base is being developed and allows surrounding tissue to behave with normal cellular immune system function.

Discussions centered on the mutuality and bidirectional effects of implant surfaces and adherent or proximal tissues. Also, the importance of state-of-the-art material and surface science to research-oriented surgeons and interactions with cellular immunologists and cell biologists was discussed. Progress in reducing infection associated with prosthesis will be rapid if multidisciplinary collaborations involving engineers on the cutting edge of surface technology can be integrated into teams of surgeons and bioengineers working on avoiding biomaterial centered infection.

Haim Bicher reviewed classical concepts of blood cell aggregation, including membrane effects, plasma factors and drug effects. He reminded us of the now recognized role of aspirin derivatives in countering platelet aggregation. He went on to describe the use of microelectrodes for monitoring oxygen tension in aggregates of cultured cells, termed spheroids, and pointed out that with cell death, oxygen tension went up in the necrotic core because of a continuing diffusion in the absence of cellular utilization. Finally, he speculated on the nature of control mechanisms for oxygen regulation in tissues in vivo, including the possibilities of an oxygen receptor.

During discussion it was emphasized that measurement of absolute oxygen tension may not be meaningful in a vital situation where there is a gradient from capillary to cell. Here the rate of transport may be important. The concept of an oxygen receptor on cells created some interest, and questions were raised as to the possibility that it could be uncoupled chemically as well as to its molecular nature. Bicher replied that it was hypothetical and that no biochemical or morphologic entity had been identified.

Tissue Engineering, pages 153–154
© 1988 Alan R. Liss, Inc.

IV. The Musculoskeletal System and Orthopaedic Surgery

- *Soft Tissue and Cartilage*

- *Bone: Mechanical, Chemical and Electrical Factors in Remodeling and Repair*

Introduction. . . . . .Van C. Mow

Over the past twenty-five years, major advances have been achieved in our understanding of the structure and functions of the musculoskeletal system. These have provided an unparalleled opportunity for the biomedical community to develop biological substitutes and replacements for injured and diseased connective tissues comprising the musculoskeletal system, particularly cartilage, bone, tendons and ligaments.

The musculoskeletal system supports loads imposed on the body, protects the vital organs and facilitates motion. Various injuries and diseases can cause acute and/or chronic failure of the musculoskeletal system. For example, arthritic destruction of a joint can cause loss of mobility or use of a limb. A bone fracture can prevent load carriage. Strains and ruptures of periarticular tissues such as tendons and ligaments can cause severe restrictions in using the hand, knee or other mobile joints. Problems related to clinical treatment modalities of these injuries and diseases stem from our incomplete understanding of the injury and repair mechanisms and etiology of the disease processes. Today, the economic impact of musculoskeletal injuries and diseases is enormous, and this is likely to increase with our aging and more active population.

Scientific and biomedical advances have led to application of engineering methods and life science principles in understanding the structure of both normal and diseased connective tissue, and to development of biological tissue substitutes to replace injured, diseased or missing tissues. The aim of such studies is to provide improved clinical treatment modalities to restore musculoskeletal function.

Areas of research include bone remodeling mechanisms and soft tissue (articular cartilage, tendon, ligament, intervertebral disc, meniscus, etc.)

maintenance and repair. Studies of the influence of chemical, electrical and mechanical stimulation on tissue homeostasis need to be undertaken. Areas of technical emphasis should include development of viable harvesting and storage procedures for cells and tissues for implantation,
biochemical and engineering characterization of ex-vivo grown cells and tissues, and development of scaled-up engineering processes for mass production of cells and tissues required for effective clinical treatment modalities.

Tissue Engineering, pages 155–160
© 1988 Alan R. Liss, Inc.

ON THE REPAIR OF MUSCULOSKELETAL SOFT TISSUES

Savio L-Y. Woo and Jennifer S. Wayne

Orthopaedic Bioengineering Laboratory,
University of California, San Diego
and San Diego Veterans Administration Medical Center

INTRODUCTION

Injuries to the musculoskeletal soft tissues such as tendons, ligaments and articular cartilage are frequent. In the general population, the incidence of seriously acute knee injuries such as anterior cruciate ligament (ACL) disruption is about 1 per 3,000 (1). Many of the injuries result in abnormal motion which leads to secondary injury of other knee structures and eventually the development of degenerative arthritis. Reconstruction procedures, i.e. tissue graft selection and fixation techniques remain arbitrary, and thus, success of clinical treatment is often considerably short of normal joint function. The digitorum perfundus tendon in zone 2 is frequently lacerated. During its repair, adhesion formation between the tendon and the fibrous digital sheath has continued to compromise the clinical outcome. Knowledge of the healing of damaged articular surfaces is also lacking because of its limited ability to repair. In the following, we will describe some of the approaches used in our laboratory for dealing with these problems.

FLEXOR TENDON

Studies have been directed to limit adhesion during healing by using improved tendon suturing techniques, by repair and reconstruction of the tendon sheath and by implementation of early motion following repair. We investigated the healing of lacerated canine flexor tendons treated with either immobilization, delayed protected mobilization or early protected mobilization using biomechanical, histological and biochemical techniques (2). At 3 week intervals through 12 weeks, the early mobilized tendons were stronger and demonstrated improved gliding function than the other postoperative groups. Scanning electron microscopy showed the gliding surface to be free of adhesion, and the DNA content at the repair site was elevated as a result of the treatment by early passive motion.

ANTERIOR CRUCIATE LIGAMENT OF THE KNEE

The forces in-vivo in the ACL are of great importance. Many studies have been designed to measure its positive and negative strains with respect to an arbitrary point (generally with the knee in full extension) (3,4,5). However, this data cannot be used directly to determine the forces in the ACL since the "zero" is not known. On the other hand, those that used "buckle type" transducers to directly measure the ACL forces in-situ (6,7) must deal with the problem of a very complex anatomy (geometry) of this tissue. An approach used by our laboratory has been to recognize that the fiber bundles of the ACL are unevenly loaded during physiological motion and external loading of the knee joint. Therefore, we have divided the ACL into discrete fiber bundles connected at distinct insertion sites on the tibia and femur.

We first measured the 6 DOF motion of the human cadaveric knee using a kinematic linkage system securely attached to the femur and tibia. With the knee at 30° flexion, external loads such as an anterior-posterior (A-P) drawer force and varus-valgus (V-V) torque were applied, and the relative motion between the two bones was recorded. The knee was then returned to the neutral position and the ACL exposed to identify and digitize each bundle's insertion site (8).

Each fiber bundle was dissected and removed from the knee together with its bony attachments. The bone-fiber bundle-bone specimen was subjected to a tensile test to failure, and the load vs. length curve for each bundle was determined (Figure 1). Based on the kinematic data, the loads in the ACL fiber bundles during A-P drawer were calculated by using these curves (Figure 2). It can be seen that the loading is not uniform between the bundles, and the anteromedial fiber bundle resisted almost all the load. Using this approach, the load in the ACL as well as its distribution during normal knee motion and with externally applied loads were determined. The extension of this method should allow the evaluation of the function of autograft, allograft, and the synthetic replacement of ACL after knee reconstruction.

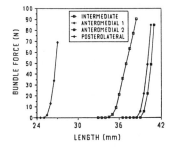

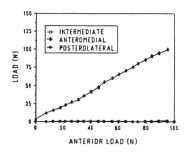

<u>Figure 1.</u> Typical load vs ACL bundle length curves from tensile tests of each bundle

<u>Figure 2.</u> Calculated loads in each ACL bundle during anterior loading of the knee

ARTICULAR CARTILAGE

One option of resurfacing damaged cartilage areas is to seek a biological solution using graft transplantation such as osteochondral allografts and perichondrial autografts. Biological tissue has the advantage of being capable of responding physiologically to joint motion. Transplantation of osteochondral grafts is clinically performed, with fresh tissue used mostly because the effects of preservation on cartilage have not been elucidated. But, there are limitations associated with fresh graft transplantation in terms of tissue procurement and time constraints, making preservation of grafts necessary to provide for a suitable selection of transplant material.

We investigated the effect of storing canine tibial plateaus in culture media (9) at 4°C on the mechanical properties of the articular cartilage (10). The biphasic theory (chapter by Mow and coworkers) was used to analyze the data obtained from confined compression creep experiments on the cartilage. An aggregate modulus and apparent permeability were determined from the analysis. Although minor changes occurred in the cartilage after the culturing process, these differences were not significant (Figure 3). The length of time in culture also did not show a significant effect by a one-way analysis of variance. Thus, the biomechanical properties measured were not significantly affected by the culturing process.

Another approach to biological resurfacing is the use of perichondrial autografts which have chondrogenic potential. Our study included an in-vivo rabbit model in which rib

perichondrium was sutured onto a cartilage denuded, bone core
from the medial condyle (11). The regenerated neocartilage at
different time points was evaluated biomechanically with a
dynamic shear tester, yielding a complex shear modulus.

The results at 100 Hz (Figure 4) indicated that the
magnitude of the complex shear modulus |G*| for rib perichon-
drium was much lower than normal articular cartilage. For the
neocartilage, the |G*| were found to lie between the normal
cartilage and perichondrium, and by 52 weeks, the |G*| for
both postoperative treatment (cage activity and continuous
passive motion) groups approached the value for normal
articular cartilage. The effect of post-operative treatment
was evident in the early healing periods, but an analysis of
variance showed that treatment modality had no overall effect
on the complex shear moduli.

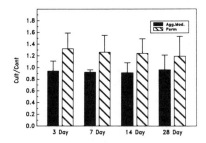

Figure 3. Normalized aggregate
modulus and permeability vs.
time in culture

Figure 4. Magnitude of
complex shear modulus for
neocartilage at 100 Hz

SUGGESTED FUTURE STUDIES (12)

The determination of tensile forces in the flexor tendon
in-vivo and the evaluation of the relative motion between the
tendon and the sheath at areas around the three interphalan-
geal joints based on kinematic analysis and experimental
studies should be done. In the region where the tendon wraps
around articular surfaces, increased compressive stresses may
be experienced. Studies on the interaction of these stresses,
and their effect on tendon healing potential must be encour-
aged. Although it appeared that immobilization of tendons
during the healing process resulted in an effective repair
without overwhelming scar formation, the signal that directs
the cells response to this activity is unknown.

Controlled passive motion has been shown to augment the healing quality of the tendon, but the optimal magnitude and duration of motion for the best healing process has not been determined. Also, an understanding of the relationship and the effect of these factors on the excursion and tension in the repair tendon is of paramount importance. Lastly, experimental studies have not addressed more complex tendon injuries, and injuries involving surrounding soft and hard tissues around the tendon as clinically relevant models will need to be investigated.

Similarly for ligaments, more information is required concerning in-vivo ligament mechanics and kinematics. The complex stress and strain distribution in different parts of the ligament under a variety of loading conditions must be established so that the normal, then repair and replacement tissues can be understood. Other innovative approaches such as the use of electricity and magnetism, ultrasound and systematic or local chemical therapy to improve the ligament and tendon repair response will need to be pursued.

One of the most intriguing areas to study is the tendon and ligament insertions to bone. The microstructure of these insertion sites and their effects on the biomechanical properties are not clearly understood. Various external factors such as immobilization and exercise have profound effects on the biomechanical properties of the insertion sites. Additional studies on the rate and mechanics which cause reversal of these deleterious effects should be pursued. Little is known about how the structure and function of insertion sites respond to overuse injury or trauma. One of the main problems in tendon/ligament reconstruction continues to be in the bony insertions, and quantitative information on the forces and force distribution at these sites will need to be investigated.

For repaired cartilage, the important topics include the relationship between the mechanical properties and the composition and organization of the repair tissue (and the degree of variability within the repair tissue), the morphological features, the structural characteristics of the molecular framework including collagen cross-linking and proteoglycans and organization of these macromolecules. The strength of bonding between the repair tissue and normal cartilage is virtually unknown. What are the influences of joint loading on the biomechanical properties of the repair tissue? What are the frictional and wear characteristics of repair tissue as compared to normal? Much engineering work remains to be done in collaboration with other disciplines to help solve these important clinical problems.

ACKNOWLEDGMENTS

Supported by the Veterans Administration and NIH grants AR 33097, AR 34264 and AR 28467.

REFERENCES

1. Daniel DM, pers comm, Kaiser Permanente Survey, 1986.
2. Woo SLY, Gelberman RH, Cobb NG, Amiel D, Lothringer K and Akeson WH.   The Importance of CPM on Flexor Tendon Healing. Acta Orthop Scand 52:615-622, 1981.
3. Arms SW, Pope MH, Johnson RJ, Ficher RA, Arvidsson I, Eriksson E. (1984). The Biomechanics of ACL Rehabilitation and Reconstruction. Am J Sports Med 12:8.
4. Hefzy MS, Grood ES, Noyes FR (1987). ACL Intra-Articular Reconstruction. Trans Ortho Res Soc 12:267.
5. Sidles JA, Larson RV, Garbini JL, Matsen FA III (1987). Ligament Length Relationships in the Moving Knee. Trans Ortho Res Soc 12:269.
6. Lewis JL, Lew WD, Schmidt J (1982). A Note on the Application and Evaluation of the Buckle Transducer for Knee Ligament Force Measurement. J Biomech E 104:125.
7. Barry D, Ahmed AM (1986). Design and Performance of a Modified Buckle Transducer for the Measurement of Ligament Tension. J Biomech E 108:149.
8. Hollis JM, Marcin JP, Horibe S, Woo SLY (1988). Load Determination in ACL Fiber Bundles under Knee Loading. Trans Ortho Res Soc 13:58.
9. Brighton CT, Shadle CA, Jiminez SA, Irwin JT, Lane, JM, Lipton, M (1979). Articular Cartilage Preservation and Storage. Arth Rheum 22:1093.
10. Wayne JS, Kwan MK, Hoover JA, Field FP, Woo SLY, Meyers MH (1988). Biphasic Mechanical Properties of Articular Cartilage from Osteochondral Shell Allografts Cultured at 4C. Trans Ortho Res Soc 13:109.
11. Woo SLY, Kwan MK, Lee TQ, Field FP, Kleiner JB and Coutts RD (1987). The Temporal Variation of Shear Modulus for Rib Perichondrium Autograft for Large Articular Cartilage Defects. Acta Orthop Scand 58:518.
12. Woo SLY, Buckwalter JA (eds) (1988). "Injury and Repair of the Musculoskeletal Soft Tissues." Illinois: AAOS.

Tissue Engineering, pages 161–166
© 1988 Alan R. Liss, Inc.

THE INFLUENCE OF MECHANICAL FACTORS ON NORMAL AND ALLOGRAFT CARTILAGE TURNOVER AND DEGRADATION.

Van C. Mow, Anthony Ratcliffe and Farshid Guilak

Orthopaedic Research Laboratory,
Departments of Orthopaedic Surgery and
Mechanical Engineering, Columbia University,
New York, NY 10032

ABSTRACT There is now a good understanding of normal articular cartilage molecular organization and structure-function relationships, but the present knowledge of cartilage repair mechanisms is limited. Biological treatment modalities using osteochondral and chondrocyte allografts offer exciting possibilities, mechanical and chemical stimuli and humoral factors can influence allograft viability and hence their success; however basic scientific data on these allografts is limited.

INTRODUCTION

Articular cartilage functions as a wear-resistant, nearly frictionless, load-bearing surface in diarthrodial joints [1]. Its composition and physico-chemical properties, as well as the ultrastructural and molecular organization of the matrix components, have profound influences on the intrinsic mechanical properties of the extracellular matrix, on the movement of fluid and the diffusional transport of substances through the tissue. These properties, in turn, enable cartilage to perform its normal biomechanical functions within diarthrodial joints. Degeneration and destruction of cartilage during osteoarthritis and rheumatoid arthritis cause enormous problems for individuals and society, in terms of patient suffering and disability, loss of work time and cost of health care: $8.5 billion annually in the USA alone. Osteoarthritis affects over 50 million people in the USA, and it is estimated that 15% of the world population suffer from some form of arthritic disease [2]. Thus basic research aimed at the understanding of normal and diseased cartilages, and the search for appropriate treatment modalities using artificial or biological substitutes are important research objectives.

This paper presents our understanding of the current research needs in developing biological substitutes for repair of cartilage and joint defects [3,4].

CARTILAGE PROPERTIES AND REPAIR

Cartilage Composition and Mechanical Properties. Articular cartilage consists of an highly organized extracellular matrix in which is embedded a sparse population of chondrocytes. These cells are responsible for the synthesis and maintenance of the macromolecules of the matrix. The major macromolecular components are Type II collagen and large aggregating proteoglycans (PG), with quantitatively minor components including other collagens (Types V, VI, IX, XI), small and nonaggregating PGs and other glycoproteins [5].

In cartilage, collagen forms a dense meshwork of fibers which gives structural shape and form to the tissue. Enmeshed in the collagen network is a high concentration of [Bs. The immobilization of PGs is enhanced by their size, their formation, into supramolecular aggregates with hyaluronate, and their ability to form networks in solution [5,6]. The PGs are predominantly of high molecular weight $(Mr\ 1-4 \times 10^6)$, and are composed of an extended protein core to which are attached many glycosaminoglycan (GAG) chains (chondroitin sulfate and keratan sulfate). These macromolecules function as carriers of stresses and strains developed within the tissue during load bearing [1,5]. The most important mechanical properties of collagen fibrils are their tensile stiffness and strength. A strong cohesive network of these fibrils is formed by intramolecular and intermolecular crosslinks between collagen monomers and collagen fibrils [5,7]. The PGs afford cartilage its compressive stiffness and swelling properties, and they maintain a high degree of hydration in the tissue [5,8,9]. Further, the high concentration of anionic groups (sulfate and carboxyl) along the GAG chains create strong intramolecular and intermolecular repulsive forces; these forces extend and stiffen the PG aggregates in the collagen network [5,8]. Mobile cations $(Na^+$ and $Ca^{2+})$ are attracted to these fixed negatively charged groups creating a large Donnan osmotic swelling pressure which acts to draw water into the tissue [9]. This swelling pressure is resisted and balanced by the tension developed in the collagen network creating a state of "pre-stress" in the collagen network.

Cartilage Biology and Turnover. Cartilage is a metabolically active tissue, with continuous turnover of its matrix components. Collagen turnover is relatively slow and it is resistant to proteolytic degradation, whereas the PGs are continually being degraded and replaced by newly synthesized PGs [5,10,11]. Studies using cartilage explants in culture showed that turnover of PGs involved release of PG fragments from the matrix into the medium, and these fragments were unable to aggregate. Structural analysis showed that the PGs released had been specifically cleaved to separate that part involved in aggregation from the major part of the molecule containing the GAG chains [10], providing an efficient mechanism of release from the matrix. In normal cartilage the chondrocytes are therefore able to maintain a balance between synthesis and breakdown.

Cartilage degradation and repair, and impairment of joint function. Cartilage degradation and loss can be regarded as the result of an imbalance of synthesis and degradation. Early events are PG loss and increased hydration in the extracellular matrix [5,8,10,11], resulting in a decrease of tissue compressive stiffness and increased permeability [8]. Changes in the tensile stiffness and strength of the collagen network appear only after significant PG depletion from the matrix [12]. Thus, disaggregation and loss of PGs, increased tissue hydration, and disruption of the collagen fibrillar network will have profound effects ability of cartilage to function as the bearing surface in diarthrodial joints.

The repair response of cartilage damage depends on the nature and extent of the injury. For example, joint immobilization, infection, and blunt trauma may cause some loss of PGs from the matrix [11,13,14]. If the chondrocytes remain viable and the collagen network remains intact, the cells can replace the lost PGs. However if the injury causes massive PG loss, disruption of the collagen network, or significant amount of chondrocyte death, then the injury will usually not be repaired and the lesions may progress to total joint degeneration. Ideally, clinical treatment would repair the injured cartilage and restore normal function and durability to the joint. As yet no method is currently available to accomplish this repair biologically.

Synthetic and biological joint replacements. Prosthetic implants and total joint replacements are currently used for repair of extensively destroyed joints; however, the relatively low life expectancy of these prostheses do not make them ideal for long term joint replacement in young patients. Improper fit, fatigue failure, and loosening are some of the most common problems associated with prosthetic joint implants. Allografting procedures may provide alternative clinical treatments of the patient requiring joint replacement or repair of joint injuries.

A current focus of treatment and repair of injured and destroyed joints is the use of allograft material. Bone allografts have proved clinically feasible with 75-80% success rates [3]. Vascularized bone and joint grafts have a decided advantage over non-vascularized grafts in their speed of healing and incorporation. However, success varies depending on the degree of genetic disparity between donor and recipient and the nature of preservation techniques applied to the graft, although freezing and freeze-drying greatly reduce osteochondral allograft immunogenicity [15].

In joint reconstruction, the osteochondral graft is probably the most common allografting procedure used and has met with varied success depending on preservation techniques, histocompatibility, and site of transplantation. While bone functions more satisfactorily as a graft material when the cells have been destroyed (due to a decrease in the immunogenic response), articular cartilage requires viable chondrocytes to survive as an allograft. Presently, there are numerous clinical methods used to promote chondrogenesis, including shaving or abrasion of fibrillated cartilage, perforation or abrasion of the subchondral bone, manipulation of joint loading and motion prior to and following injury, laser stimulation, and extraction of PGs or suppression of PG synthesis. Recent clinical investigations show promising results using a variety of grafted materials including osteochondral fragments, periosteum, perichondrium, and chondrogenesis stimulating factors such as various growth hormones or bone morphogenic proteins [3,15,16]. However, long term viability of the cartilage from these allografts remains to be determined. Indeed, no attempts have been made to assess the biology, biochemistry and biomechanics of these types of graft materials.

Without treatment, some full-thickness defects of articular cartilage can undergo repair. Although this appears to be satisfactory histologically, the chemical composition of the reparative tissue may contain significant amounts of Type I as well as Type II collagen. Thus, the structure of the reparative tissue is imperfect and results in fibrillation and eventual degeneration [17]. It is therefore vital that the phenotypic expression of the chondrocytes maintains identical or very similar to that of chondrocytes in normal articular cartilage.

A recent and promising approach to this problem has been to introduce a new cell population to repair a cartilage defect by the implantation of isolated chondrocytes immobilized in a gel, allowing their subsequent synthesis of a new cartilage matrix. Recent studies have shown that cultured homologous embryonic chondrocytes embedded in a biologically resorbable gel placed in a surgically created defect on the joint surface can repair the injury for up to 18 months [4]. No detailed biological, biochemical and biomechanical investigation has ever been pursued to study this exciting possibility. This technique offers a new avenue of research in developing biological substitutes for joint repair.

The Mechanical Stimulation and Modulation of Cartilage Composition and Structure. It has been shown in vivo and in vitro that the mechanical environment around cartilage has a significant effect on the metabolic rate of the chondrocytes and the biochemical composition of the tissue [11, 14]. In disuse, the chondrocyte becomes almost quiescent and this is accompanied by a decrease of cartilage thickness and PG content; remobilization results in complete reversal of these atrophic changes. Recent in vitro experiments have shown that cartilage allografts cultured under cyclic load exhibited superior histological and biomechanical properties when compared to tissue which was cultured but not loaded [11,14]. However, no systematic study has quantified the dose-response of cartilage stimulated by mechanical and biochemical means, although this may have immediate clinical significance.

FUTURE DIRECTIONS

Recent advances indicate that repair of articular cartilage defects may be engineered with appropriate choice of biomechanical, biochemical, and biological parameters. Many avenues are open for the exploration of structure-function relationships, degradation and repair, and interactions of subcomponents within cartilage. Important future studies should include the following topics:

1. The relationship between the intrinsic mechanical properties of repair cartilage and its composition and molecular organization.
2. The dose-response of mechanical stimuli on turnover, composition, and material properties of cartilage.
3. The biomechanical and biochemical characterization of the normal tissue/repair tissue interface.
4. The control of cartilage regeneration by maintenance of chondrocyte phenotype via mechanical stimulation or growth factors and pharmacological agents.
5. The tissue-matching, storage and long term viability of osteochondral and chondrocyte allografts.

REFERENCES

1. Mow, VC and Mak, AF: In Handbook of Bioengineering. New York, McGraw-Hill, pp. 5.1-5.34, 1986.
2. Grazier, KL et al.: The Frequency of Occurrence, Impact and Cost of Selected Musculoskeletal Conditions in the United States, AAOS Press, Chicago, 1984.
3. Friedlaender, GE and Mankin, HJ: In Ann. Rev. Med., pp. 311-324, 1984.
4. Itay, S, et al.: Clin. Orthop., 220: 284, 1987.
5. Muir, H: Biochem. Soc. Trans., 11:613, 1983.
6. Hardingham, TE, et al.: J. Orthop. Res. 5:36, 1987.
7. Woo, SLY, et al.: In Handbook of Bioengineering. New York, McGraw-Hill, pp. 4.1-4.44, 1986.
8. Mow, VC, et al.: J. Biomech., 17:377, 1984.
9. Maroudas, A: In Adult Articular Cartilage. 2nd ed. Ed. M Freeman. Tunbridge Wells, England, pp. 215-290, 1979.
10. Ratcliffe, A, et al.: Biochem. J., 238:571, 1986.
11. Tammi, M et al.: In Joint Loading: Biology and Health of Articular Structures, ed. HJ Helminen, John Wright, Bristol, pp. 64-88, 1987.
12. Akizuki, S, et al.: J. Orthop. Res., 4:379, 1986.
13. Donohue, JM, et al.: J. Bone Jt Surg., 65A:948, 1983.
14. Slowman, SD and Brandt, KD: Arth. Rheum., 29:88, 1986.
15. Czitrom, AA, et al.: Clin. Orthop., 206:141, 1986.
16. O'Driscoll, S, et al.: J. Bone Jt Surg., 68-A, 1986.
17. Buckwalter, JA et al.: In: Injury and Repair of Soft Tissues, AAOS Press, Chicago, pp 465-482, 1988.

Tissue Engineering, pages 167–172
© 1988 Alan R. Liss, Inc.

MODELING SOFT TISSUE BEHAVIOR[1]

Allen H. Hoffman, Peter Grigg,[*]
Holly K. Ault & David M. Flynn

Mechanical Engineering, Worcester Polytechnic Institute
Worcester, MA   01609

[*]Dept. of Physiology, Univ. of Massachusetts Medical School
Worcester, MA   01605

ABSTRACT   A series of interrelated experimental and modeling activities have been undertaken as part of a continuing study of mechanoreceptors located in the posterior knee capsule of the cat.   Specific accomplishments include: development of finite element based methods to measure strains in soft tissue, development of biomechanical models for the calibration of mechanoreceptors as *in vivo* force transducers, determination of anisotropic material properties and the development of improved micromechanical models based upon the tissue structure.

INTRODUCTION

The nature of soft tissues make them difficult to study. Soft tissues are very compliant and undergo large strains when loaded.   These tissues are nonlinear, nonhomogeneous, anisotropic and viscoelastic. *In vivo,* soft tissues generally are under tension.   The complexities of soft tissue behavior have prevented the development of accurate constitutive equations in all but the simplest of loading and geometric situations.   Micromechanical models developed to bridge the gap between microscopic structure and macroscopic behavior have also met with limited success.

For the past several years our laboratory has been investigating mechanically sensitive neurons innervating the posterior knee capsule of the cat.   These studies have led

---

[1]This work was supported by NIH grant NS10783.

to detailed investigations of the joint capsule tissue. The difficulties inherent in developing an accurate constitutive equation to predict strain under loading led to the development of finite element based methods to experimentally measure strains in soft tissue (1). The anatomical structure of the posterior knee capsule of the cat led to the development and verification of a biomechanical model of the upper margin of the capsule as a catenary suspension cable (2). Using the model, a method was developed to calibrate individual mechanoreceptors residing in the cable as *in vivo* force transducers (3). The need to understand local material properties in a nonhomogeneous tissue has led to the development of finite element based methods for determining the anisotropic material constants (4). Finally, since the macroscopic behavior of soft tissue is based upon its microscopic structure and properties of its constituent materials, it has become necessary to develop improved micromechanical models. The following sections describe our studies in each of these areas.

MEASURING STRAINS IN SOFT TISSUE

If a tissue is excised and studied in isolation, then the *in vivo* loading state is lost and the exact boundary conditions can not be duplicated. We developed a method for measuring the *in vivo* components of plane strain in a soft tissue based upon a grid-like array of markers on the tissue surface (1). Four adjacent markers are treated as the nodes of a four node isoparametric finite element. Marker displacements are measured directly using a high sensitivity television camera and a frame grabbing arrangement. Finite element method mathematics are then used to calculate the large strain tensor at any point within the element. The method has been implemented using non-rectangular quadrilateral elements that are approximately 2 mm on each side. Internal soft tissue membranes have been analyzed using radiopaque markers. Recently Humphrey et al (5) have reported a real-time use of the method.

BIOMECHANICAL MODELS

The anatomical arrangement of some soft tissue structures lend themselves to the direct development of a biomechanical model. For example, the upper ligamentous margin of posterior knee capsule in the cat and other small mammals can be modeled as a catenary suspension cable from which the capsule sheet is suspended (2). This model was developed after visually observing the structure, studying its morphology using scanning electron microscopy and measuring the material properties of the tissue. These studies revealed the upper margin to be a one dimensional structure attached to sesamoid bones at each end. This model allowed us to develop a method for calibrating the response of mechanoreceptors residing in the cable in terms of the local cable tension (3). Using calibrated mechanoreceptors, we have been able to study capsule loading as the joint is rotated into extension and more recently under the condition of rotation about the long axis of the tibia.

The usefulness of any biomechanical model rests upon how well it predicts the actual tissue behavior. Usually models are developed to meet specific needs and are restricted in their range of application. All tissues can not be effectively modeled. As our example indicates, useful models need not be highly sophisticated.

DETERMINATION OF ANISOTROPIC MATERIAL PROPERTIES

In quasi-static applications the viscoelastic effects are minimal and the tissue can be considered to be pseudoelastic. Since soft tissues are nonhomogeneous, the determination of local (anisotropic) material properties is of interest. A finite element based method, applicable to soft tissues, has been developed to determine the material properties of planar sheets and membranes (4). The method is based upon a four node quadrilateral element and can be used to determine the local properties of materials which range from isotropic to anisotropic. The method is based upon the equation

$$\mathbf{F} = \mathbf{K}\,\mathbf{U} \qquad\qquad\qquad (1)$$

where $\mathbf{F}$ is the nodal force vector, $\mathbf{U}$ the is nodal displacement vector and $\mathbf{K}$ is the stiffness matrix which contains the material properties. If $\mathbf{F}$ and $\mathbf{U}$ are known then the material properties contained in $\mathbf{K}$ can be found. The method requires that the tissue be excised and placed in a load frame that applies known nodal forces. Nodal displacements are measured from frame grabbed images. The tangent properties of nonlinear materials can be determined by applying a series of incremental loadings about a reference state. Up to four independent nodal load and displacement sets must be developed experimentally. Large strain and large displacement theory have been included to account for geometric nonlinearities. At present the method has only been applied to soft tissues using large element sizes (1 x 2 cm). We are actively working on refining our techniques to reduce the element size and thereby determine more localized properties.

MODELING DYNAMIC SOFT TISSUE BEHAVIOR

In many applications the dynamic behavior of soft tissues is of interest and the viscoelastic behavior of the tissue becomes important. From a mechanics standpoint, this greatly increases the complexity of analyzing the material behavior. Fung (6) discusses the analytical methods that are required. Due to the complexity of the method, few soft tissues have been effectively analyzed.

DEVELOPMENT OF MICROMECHANICAL MODELS

The macroscopic behavior of tissues is controlled by their microscopic structure. A number of micromechanical models of soft tissue have been developed. Some models are based on beam theory (7,8) or equilibrium analyses (9), others use composite theories (10,11). The former assume that the main load bearing component within the material is collagen, neglecting contributions of the matrix. However, significant changes in material properties of tissues have been correlated to presence of matrix (12). Composite micromechanical models rely on knowledge of the mechanical properties of the constituents, their geometrical config- uration, and interactions between the components. There is considerable technical uncertainty in all of these areas. As a result, present models may assume values for material properties, use simplified geometries, or use 1- or 2-D models. We focus on two areas: 1) characterization of the geometric structure of collagen fibers in the tissue and 2) development of improved 3-D composite type micromechanical models for structures under quasi-static loading.

Micromechanical models of tissue rely upon histological information regarding the orientation of collagen fibers within the tissue. The orientation of fibers may be described analytically by a distribution function. A method has been developed to determine the 2-D distribution of fiber orientations from SEM images of connective tissue using a grey scale gradient analysis technique (13). An IBM

PC frame grabs the SEM images and convolves them with a gradient operator. The direction of the grey scale gradient is normal to the fiber edges. Local fiber directions are found and collected into a histogram representing the desired distribution function, which is used as input to the micromechanical model.

Many promising models for soft tissue behavior can be found in the composite materials literature. In classical composite models for both fibrous and particulate composites, material properties of the components are combined using rules based on the geometric configuration of and interactions between the components. A critical decision in any composite model is the manner in which the component properties are aggregated.

Bounds on elastic properties of composites are found using the theorems of minimum potential energy and least work. With the former, we assume a series coupling of the components as the rule for aggregation. With a constant stress field, this is equivalent to adding compliances of the components. A lower bound on the elastic properties is obtained. For the upper bound, we assume that the components are coupled in parallel. With this model under constant strain, we add stiffnesses. For the same geometric configuration, the composite properties predicted by these models is quite different. Improved models require more information on the stress and strain states within the material, the nature of interactions between constituents, and additional geometric data. Bounds predicted by improved models may tend to converge, as they do for some simple geometries used in classical composite models.

In the near term, micromechanical models will be improved by creating better methods for developing geometric data from SEM micrographs and improved model aggregation methods. Large scale improvements of micromechanical models await increased understanding of the microstructure of these tissues, including knowledge of the effects of histological processes (eg. dehydration, fixation) on the internal geometry of the material. We also require improved understanding of mechanical properties of the constituents of soft tissue and their interactions.

SUMMARY

Effective soft tissue modeling is based upon a synthesis of experimental and analytical methods. Experimental techniques based upon the mathematics of finite element methods have been developed to measure strains in soft tissue and to determine anisotropic material properties. In some anatomic situations, effective biomechanical models can be directly developed. Micromechanical models of soft tissue hold great promise, but at present there is considerable uncertainty in the values of the parameters which serve as input to these models.

REFERENCES

1. Hoffman AH, Grigg P (1984). A method for measuring strains in soft tissue. J Biomech 17:795.
2. Hoffman AH, Grigg P, Flynn DM (1985). A biomechanical model of the posterior knee capsule of the cat. J Biomech Eng 107:140.
3. Grigg P, Hoffman AH (1988). Calibrating joint capsule mechanoreceptors as in vivo soft tissue load cells. J Biomech (submitted).
4. Flynn DM, Hoffman AH, Grigg P (1988). Finite element model for determining material properties. Proc 14th Northeast Bioeng Conf. IEEE (in press).
5. Humphrey JD, Vawter DL, Vito RP (1987). Quantification of strains in biaxially tested soft tissue. J Biomech 20:59.
6. Fung YC (1981). "Biomechanics: Mechanical Properties of Living Tissue." New York: Springer-Verlag, p 41.
7. Diamant J, Keller A, Baer E, Litt M, Arridge RGC (1972). Collagen: Ultrastructure and its relation to mechanical properties as function of aging. Proc R Soc Lond 180:293.
8. Comninou M, Yannas IV (1976). Dependence of stress-strain nonlinearity of connective tissues on the geometry of collagen fibers. J Biomech 9:427.
9. Lanir Y (1979). A structural theory for the homogeneous biaxial stress-strain relationships in flat collagenous tissues. J Biomech 12:423.
10. Lanir Y (1978). Structure-strength relations in mammalian tendon. Biophys J 24:541.
11. Lanir Y (1983). Constitutive equations for fibrous connective tissues. J Biomech 16:1.
12. Minns RJ, Soden PD, Jackson DS (1973). The role of the fibrous components and ground substance in the mechanical properties of biological tissues: A preliminary investigation. J Biomech 6:153.
13. Ault HK, Hoffman AH, Grigg P (1988). A method for determining fiber direction in soft tissues. Proc 14th Northeast Bioeng Conf. IEEE (in press).

Tissue Engineering, pages 173–179
© 1988 Alan R. Liss, Inc.

THE REGULATION OF SKELETAL BIOLOGY
BY MECHANICAL STRESS HISTORIES

Dennis R. Carter

Department of Mechanical Engineering
Stanford University
Stanford, CA 94305 U. S. A.
and
Rehabilitation Research and Development Center,
Veterans Administration Medical Center,
Palo Alto, CA 94304 U. S. A.

ABSTRACT   The growth, maturation and aging of the skeleton is accomplished by the proliferation, maturation, degeneration, and ossification of cartilage.   Bone tissue which replaces cartilage during maturation is modeled and aligned to provide apparent density and trabecular orientations to efficiently resist the loads to which it is exposed.   Osteoarthrosis in the aged involves the degeneration and destruction of the remaining cartilage on the joint surface and osteophyte formation.   Skeletal tissue regeneration in, for example, fracture healing or joint repair is accompanied by tissue differentiation processes which are similar to those observed during development.   All of these biological events are influenced by the mechanical forces to which the skeleton is exposed.   A general theory for the mechanical regulation of biological processes in the chondro-osseous skeleton is presented here which is based on the loading history of the tissue. This theory can be used to describe and predict the biological events and morphology of long bones throughout life, beginning the with the embryonic cartilage anlagen.

INTRODUCTION

The chondro-osseous skeletal system is the structural framework of the body and provides its characteristic form.   Bone, cartilage, and associated fibrous tissues are subjected to repeated stresses and strains during millions of cycles of mechanical loading throughout life.   The repeated loading over a specific period of time constitutes the local stress (or strain) histories of the skeletal tissues.   It is now acknowledged by virtually all investigators that mechanical loading history has a significant influence on skeletal biology.

Our research group has proposed that the natural sequence of events in all cartilage in the appendicular skeleton is proliferation, maturation, degeneration and ossification (1, 2). The application of intermittent shear (deviatoric) stress (or strain energy) into cartilage will accelerate and intermittent compressive hydrostatic (dilatational) stress will retard or arrest these processes. The very existence of articular cartilage in a mature human diarthrodial joint can thus be ascribed to the absence of significant shear stress and the inability of the subchondral ossification front to proceed into areas of high magnitude hydrostatic pressure. These mechanical principles also appear to govern or strongly influence the creation of growth plates and regulate the position and shape of these structures. As ossification occurs in the developing anlage, the new bone immediately adapts its apparent density and orientation to the local stress histories. The articular cartilage thickness distributions in normal joints and the characteristic cartilage and bone changes in arthrosis are consistent with the stress history principles which guide endochondral ossification in morphogenesis. In this sense, degenerative joint disease may be viewed as the final stage in the ossification of the cartilage anlage. Regeneration, mesenchymal cell differentiation and repair of skeletal tissues are also directed by mechanical stress in a manner consistent with skeletal morphogenesis. This presentation will review some of the recent research results which have lent support for this view of the role of mechanical stress histories in chondro-osseous biology.

MORPHOGENIC ENDOCHONDRAL OSSIFICATION

A plane strain finite element model was created to represent the proximal half of the femur from 48 days after conception to a postnatal age of approximately 18 months (1). The loading history imposed on the proximal femur was represented by three loading conditions which were assumed to be intermittently applied for an equal number of loading cycle over a specific time period. Initially, the entire model was composed of cartilage. At five stages of development, the stress fields were solved for the three loading conditions. The stimulus to endochondral ossification in each cartilage element was represented by adding the strain energy density values calculated under these different loading conditions. Since cartilage was modeled as a nearly incompressible, linearly elastic material, the strain energy density reflects primarily the shear (or deviatoric) energy. The cumulative, superimposed strain energy densities calculated were displayed as contour plots.

For the all-cartilage model, the highest superimposed strain energy density was on the periosteal surface at the midshaft of the diaphysis, corresponding to the formation of the tubular diaphysis. As ossification moves toward the bone ends, the strain energy density in the cartilage immediately ahead of the ossification front was found to be higher than average and was greatest at the periosteal surface. In what will be the metaphyseal area, the highest strain energy was still adjacent to the mineralization front. The strain energy, however, was not as concentrated at the periosteum but was distributed more in the interior regions where cancellous bone eventually forms. When the mineralization front approached the epiphyseal region, the energy distributions underwent a distinct change. An area of high strain energy was produced in the center of the chondroepiphysis where the secondary ossific nucleus appears.

At all developmental stages, triaxial compressive stresses were calculated near the joint surface directly under the applied contact stress. This stress state reflects the imposition of high magnitude, intermittent hydrostatic pressure and relatively low strain energy density (and shear stress) in the area which becomes the articular cartilage. After the osseous epiphysis forms, the articular cartilage is still exposed to intermittent hydrostatic pressure which prevents the degeneration and ossification of the cartilage. The magnitudes, depth of penetration, and frequency of these hydrostatic pressures, in balance with the cyclic shear stresses, control the cartilage thickness in a mature joint.

In related studies, we have demonstrated how the loading history can be incorporated to predict diarthrodial joint morphogenesis (3, 4) and the growth and development abnormalities associated with congenital dislocation of the hip (5). We have also used our loading history techniques to predict ossification patterns in the developing human sternum (6). By subtly varying the geometry of the model, we were able to duplicate the three standard sternal ossification patterns.

DEGENERATIVE JOINT DISEASE

The osteoarthrotic changes in cartilage and subchondral bone are so closely related that it may be difficult to identify a unique initial event in the disease process. There is a very slow, continuous growth and remodeling process at the articular ends of bones throughout life. This site of cartilage ossification is, in fact, the arrested ossification front which was instrumental in the formation of the osseous epiphysis during skeletal development. In arthrosis, there are significant changes in both the cartilage and bone and this remodeling process is altered.

Carter et al. (7) calculated the mechanical deformations in the subchondral bone of the femoral head and acetabulum. The magnitude of the subchondral bone compressive hydrostatic stress was found to correlate with cartilage thickness and was highest in the superior femoral head and moderate at the acetabular roof. Areas of high surface contact pressures on the femoral head were also areas of high hydrostatic compression in subchondral bone and cartilage and rarely the initial sites of degenerative change. The seldom-contacting surfaces of the medial-inferior and peripheral areas of the femoral head and the roof of the acetabulum were areas which combined a low hydrostatic compressive component with subchondral bone tensile strains (and stresses) tangential to the joint surface. Initial cartilage fibrillation and osteophyte formation are often found in these areas. The findings suggest that fluctuating hydrostatic pressure in a functional joint inhibits vascular invasion and the degeneration and ossification of articular cartilage. The generation of tensile strain may accelerate the osteoarthrotic process by direct mechanical damage. Additionally, since tangentially oriented tensile strains are associated with a reduction in the compressive dilatational stresses and an increase in shear stresses, they may permit or promote vascular invasion, cartilage degeneration, and osteophyte formation.

These mechanical principles in osteoarthrosis are the same as those which have been previously demonstrated to guide the ossification of the cartilage anlage during skeletal morphogenesis and thereby effectively link the degenerative process to growth and development .

BONE ARCHITECTURE AND MAINTENANCE

Repeated mechanical loading on the cartilage anlagen continues to be applied during and after ossification and, indeed, throughout life. Carter et al. (8) characterized the bone loading histories over some period of time in terms of stress magnitudes or cyclic strain energy density and the number of loading cycles. This loading history approach can be applied to predict variations in apparent density within and among bones (2, 9).

The stress history model for bone has been applied to successfully predict the distribution of apparent density and trabecular orientation in the adult proximal femur using finite element models (2). The results indicated that the hollow diaphyseal shaft and the trabecular morphology of the femur can only be explained by considering the cumulative influence of the joint loadings from multiple directions. We believe, based on these results and similar results with the tibia, that the entire internal structure, the existence of a medullary cavity and distributions of trabecular morphology in long bones are governed primarily by the full loading histories to which the bone is exposed during growth and throughout life. We suggest that three dimensional finite element models with carefully defined loading histories would be successful in predicting bone morphology in long bones throughout the appendicular skeleton.

TISSUE DIFFERENTIATION AND FRACTURE HEALING

Fracture healing is a regenerative process which displays many of the tissue differentiation characteristics observed during skeletal morphogenesis. In both skeletal morphogenesis and fracture healing, tissue differentiation is believed to be sensitive to blood supply and tissue oxygen tension. Osteoprogenitor precursor cells in regions of poor vascularity or low oxygen tension tend to be shunted into a chondrogenic rather than osteogenic pathway. Mechanical compression is thought by some to have a similar influence in favoring chondrogenesis.

We have constructed a semi-quantitative hypothesis for tissue differentiation in fracture healing which is consistent with the theories of morphogenesis previously presented (10). The intermittent stresses and the vascular supply to undifferentiated mesenchymal cells present in the early callus are assumed to be the major local factors to be considered. A schematic view of our perspective of tissue development under cyclic loading of undifferentiated mesenchymal tissue is shown in Fig. 1. In considering this figure, one must keep in mind that the entire loading history, over some period of time, will determine the tissue response. The creation of just a few "bad" cycles with high shear or tensile hydrostatic stresses may be sufficient to promote fibrous tissue formation at the expense of osteogenesis. One must also be cognizant of the fact that the tissue under consideration is relatively compliant. The stress magnitudes are therefore very small compared to what may be experienced by, for example, bone tissue. We proposed that in this early callus material: 1) Fracture elicits an osteogenic stimulus. 2) If minimal cyclic stresses (or strains) are created and there is a good blood supply, bone will form directly. 3) High stress magnitudes will encourage tissue proliferation. 4) High shear and/or tensile hydrostatic stresses encourage fibrous tissue formation. 5) High compressive hydrostatic stresses encourage chondrogenesis. 6) If cartilage or fibrocartilage forms, cyclic shear will promote and compressive hydrostatic stresses will inhibit endochondral ossification.

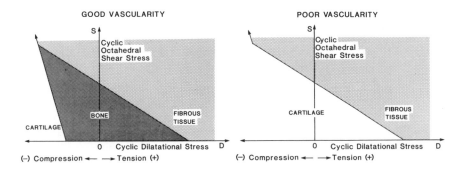

Figure 1: Tissue Differentiation caused by cyclic stresses

The application of high interrmittent shear and/or hydrostatic tensile stress will cause fibrous tissue formation. If this loading history is continued, it is unlikely that this tissue will ossify. If cartilage or fibrocartilage is formed, there is the possibility that endochondral ossification will occur. We hypothesized that the influence of mechanical stress histories in callus ossification is similar to that during skeletal morphogenesis; cyclic compressive hydrostatic stresses inhibit or prevent ossification and cyclic shear stresses encourage proliferation and ossification.

To test our tissue differentiation concepts, two-dimensional finite element models of a healing osteotomy in a long bone were generated and the theory was used to predict the regional variation in the early callus tissue characterizations and the locations of initial ossification. The analysis approach used in relating the loading history to tissue differentiation was similar to that used for investigating endochondral ossification during joint morphogenesis (4).

Our predictions for tissue differentiation and initial bone formation appear to best match previous experimental studies and clinical observations when we assumed that the fracture healing patterns are more strongly controlled by the intermittent hydrostatic pressure (dilatational stress) in the initial callus than by the intermittent shear stress.   Our models were then consistent with the prediction of chondrogenesis in the interfragmentary gap and more fibrous tissue and perhaps direct intermembranous bone in other callus regions.   This prediction is corroborated by the histological and histochemical findings of other investigators.   The strongest stimulus for bone formation is directly outside of the interfragmentary gap ant the periosteal (and endosteal) surface.   This is an area where high compressive dilatational stresses are absent.

The apparently dominating role of intermittent hydrostatic stress in determining tissue differentiation in the initial callus may be due to the role of intermittent hydrostatic pressure in the inhibition of vascular ingrowth.   In our fracture healing analyses, the pattern of hydrostatic (dilatational) stresses correspond to the angiogenesis and ossification patterns reported by others.   Revascularization and ossification appear first in those areas which do not experience high compressive hydrostatic stresses.   We have demonstrated that the pattern of vascular morphogenesis in the femoral head and neck of the cartilage anlage may also be influenced by the intermittent hydrostatic stress distributions (1).

SUMMARY

We suggest that phenotypic expression during growth, development, and aging is regulated by mechanical stress histories to a much greater extent than has previously been thought.   The versatility and power of the theory presented here suggests that it represents fundamental framework for understanding the role of mechanical stress in all chondro-osseous biological processes.   Application of this theory in the study of fracture healing and non-union, synovial joint neochondrogenesis, growth plate biology, congenital skeletal disorders, bone remodeling in the response to prosthetic implants, the influence of exercise on bone mass, and other areas of regeneration and repair are appropriate.

ACKNOWLEDGEMENT

Supported in part by NIH grant AR 32377 and the Veterans Administration Rehabilitation Research and Development Program.

REFERENCES

1.   Carter DR, Orr TE, Fyhrie DP, and Schurman DJ, (1987) Influences of mechanical stress on prenatal and postnatal skeletal development. Clin Orthop & Rel. Res 219:237-250.

2.   Carter DR, (1987) Mechanical loading history and skeletal biology. J Biomechanics 20:1095-1109.

3. Carter D.R. and Wong M, (1988) Mechanical stresses and endochondral ossification in the chondroepiphysis. J Orthop Res 6:148-154.

4. Carter DR and Wong M, (1988 in press) The role of mechanical stress histories in the development of diarthrodial joints. J Orthop Res 6.

5. Orr TE and Carter DR, (1988) Mechanical stresses and developmental abnormalities in congenitally dislocated hips. Trans Orthop Res Soc 34th Annual Mtg 13:234.

6. Wong M and Carter DR, (1988) Mechanical stresses and morphogenic endochondral ossification in the sternum. Trans Orthop Res Soc 34th Annual Mtg 13:241.

7. Carter DR, Rapperport DJ, Fyhrie DP, and Schurman DP, (1987 in press) Relation of coxarthrosis to stresses and morphogenesis: A finite element study. Acta Orthop Scand 58.

8. Carter DR, Fyhrie DP, and Whalen RT, (1987) Trabecular bone density and loading history: Regulation of connective tissue biology by mechanical energy. J Biomechanics 20:785-794.

9. Whalen RT, Carter DR, and Steele CR, (1987) The relationship between physical activity and bone density. Trans Orthop Res Soc, 33rd Annual Mtg, 12:463.

10. Carter DR, Blenman PR, and Beaupre GS, (1988 in press) Correlations between mechanical stress history and tissue differentiation in initial fracture healing. J Orthop Res 6.

Tissue Engineering, pages 181–187
© 1988 Alan R. Liss, Inc.

STRAIN ASSESSMENT BY BONE CELLS[1]

Stephen C. Cowin

Department of Biomedical Engineering,
Tulane University, New Orleans, Louisiana 70118

ABSTRACT The potential mechanisms by which
bone cells sense strains and subsequently
model or remodel to adjust to these strains is
reviewed. A model for the surface deposition
and resorption of bone tissue developed from
the cellular level concepts is reviewed.

INTRODUCTION

The transduction of mechanical information
to chemical information, the mechanism underlying
the functional adaptation of bone tissue (Wolff's
Law) has not been established. Furthermore, the
means by which cell populations are subsequently
controlled has not been identified. In this
contribution the evidence concerning the biological
strain transduction mechanism is briefly reviewed.
A model for the surface deposition and resorption
of bone tissue based on the cellular level concept
of a cell activity function is reviewed. The
applications of this model to the prediction of
bone modeling and remodeling are briefly
summarized.

[1]This investigation was supported by USPHS,
Research Grand DE06859 from the National Institute
of Dental Research, National Institutes of Health,
Bethesda, MD 20205

POTENTIAL STRAIN TRANSDUCTION MECHANISMS

Strain

The normal or axial strain is denoted by ε, and it is defined as the ratio of the change ΔL in a reference or gage length to the original reference or gage length L, $\varepsilon = \Delta L/L$. The change in the gage length is given by $\Delta L = \varepsilon L$. A typical physiological strain level is $\varepsilon = 0.002$. If the gage length is the bone cell size of 10μ, then $\Delta L = 2nm = 20\text{Å}$. If the gage length is 2cm, then $\Delta L = 40\mu$ or about $400,000$ Å.

Bone Cells

Menton et al. (1) use the SEM to present a pictorial history of the bone deposition process. They note that there is general agreement that all periosteal and endosteal bone surfaces are completely covered with bone lining cells (BLC). These cells are packed in a polygonal manner on the bone surface. Among the bone lining cells are undifferentiated osteoblasts or osteoblastic precursor cells. As a BLC becomes an osteoblast and matures, it develops numerous complexly branched processes that contact the processes of maturing and mature osteocytes. As the mineralization front approaches the osteoblast, the osteoblast appears to lose its processes, it becomes encased in a lacuna, and begins to function as an osteocyte.

There appear to be two bone cell lineages, one for the osteoblast/osteocyte lineage and one for the osteoclastic lineage. Chambers (2) uses the term immigrant to describe the osteoclastic lineage because osteoclasts take origin from a cell which can reach the bone via the circulation. He notes that it is not clear that mononuclear phagocytes are the most plausible precursors of osteoclasts. The overall activity of osteoclasts is regulated in accordance with the role of bone as a reservoir of mineral for plasma calcium homeostasis; local osteoclastec activity is determined by bone modeling and remodeling. Chambers (2) presents a working hypothesis for the local and systemic regulation of bone resorption. The hypothesis assumes that strain stimulates the osteocyte in the lacunae of the mineralized bone tissue and that the osteocyte transmits this information to the BLC's. The BLC's then induce osteoblasts to produce osteokinetic agents which tell passing osteoclasts where to localize. Osteoblasts thus have the capacity either to suppress (through prostaglandin production) or to stimulate (through mineral exposure) osteoclastic resorption.

Bone Cell Strain Sensitivity

For more than a decade Lanyon et al. (3) have been evaluating the relationship between bone tissue response and strain magnitude. They have shown that bone resorption will occur for strains less than about 0.001 and bone deposition will occur for strains greater than about 0.003. Between these two strain limits bone tissue appears to enjoy remodeling equilibrium. The bone cell displacements ΔL associated with the strains $\varepsilon = 0.001$ and $\varepsilon = 0.003$ are $10\mu(0.001) = 1nm = 10\text{Å}$ and $10\mu(0.003) = 3nm = 30\text{Å}$. If bone cells are the resorption-deposition transducers, they must be able to sense displacements of less than 10Å and emit resorption signals and sense displacements greater than 30Å and emit deposition signals. It has not been shown that bone cells have this degree of strain sensitivity.

Streaming Potentials

Recent studies of the electrical effects in bone suggest that electrical potentials of electrokinetic origin dominate over those of piezoelectric origin in fluid filled bone, Gross and Williams (4), Pollack, et al. (5). There are many fluid filled cavities in bone tissue, including lacunae, canaliculi, Haversian and Volkmann canals. When a bone is cyclically strained, the fluid in these cavities flows past the solid surface forming the boundary of the cavity. In bone tissue the fluid is charged and the charge is

convected with the fluid giving rise to a streaming
current which, in turn gives rise to a strain
generated potential (SGP). The magnitude of the SGP
is sufficient to be osteogenic.

Biological Strain Memory Speculation

In vivo experiments by Skerry, et al. (6) have
demonstrated that the orientation of proteoglycan
(PG) molecules within bone tissue is affected by
intermittent loading, at physiological strain
magnitudes. In the absence of further loading, the
loading related effect on PG orientations persists
for over a day. A direct physical connection between
PG and many types of bone cells has been established
(7). These results suggest the possibility of a
mechanism in bone that remembers strains and the
orientation of the applied strains.

SURFACE DEPOSITION-RESORPTION MODEL

In this section the elements of a model
describing the net osteoblastic and osteoclastic
activity on the surface of mineralized bone tissue is
described. The surface can be that of a trabecula,
Haversian canal or the periosteal or endosteal
surface of a whole bone.

Bone Cell Activity Function

Bone cell activity is assumed to depend upon
genetic, metabolic and hormonal factors as well as
the strain history experienced by the bone cell. The
activity function is written as $a_b(\varepsilon)$ or $a_c(\varepsilon)$ to
differentiate between the activity functions for
osteoblasts and osteoclasts. Experimental studies
related to the measurement of $a_b(\varepsilon)$ include those of
Rodan et al. (8), Nulend (9), and Gross et al (10).

Model Formulation

Let U, with the dimensions of velocity, denote
the net surface deposition or resorption. Typical

velocities would be measured in units of microns per day or mm's per year. Let n_b and n_c denote the number of osteoblasts and osteoclasts, respectively, per unit area. Let A_b and A_c denote the surface area available to the osteoblasts and osteoclasts, respectively. Following the ideas of Martin (11) and Hart (12), U is expressed in terms of these parameters by the following expression:

$$U = n_b A_b a_b - n_c A_c a_c .$$ [1]

Remodeling (modeling) equilibrium is the term used to describe the situation when U is zero. In this case osteoblastic and osteoclastic processes have a zero net effect.

There are many possible forms for the dependence of the cell activity functions on strain history, the simplest occuring when the activity functions a_b and a_c have a linear dependence on the strain ε,

$$a_b = H_b \varepsilon + G_b , \quad a_c = H_c \varepsilon + G_c,$$ [2]

where H_b, G_b, H_c and G_c are constants. When equations [1] and [2] are combined it is possible to express U in the form

$$U = C(\varepsilon - \varepsilon_o)$$ [3]

where C and ε_o are constants. Equation [3] represents the hypothesis of Cowin and Van Buskirk (13) for surface deposition and resorption of bone tissue. The constant ε_o represents a strain at which no remodeling occurs and the constant C is the remodeling rate coefficient.

The Total Model

A complete model for the prediction of bone tissue adaptation to applied strains employs traditional mechanical models in conjunction with the relationship [3]. The traditional mechanical model is the theory of elastic bodies based on Hooke's law of proportionality between stress and strain and Newton's three laws. The complete theory is a modification of the theory of elastic bodies in that the surfaces of the bodies move according to the rule described by equation [3]. As the bone evolves to a new shape, the stress and strain in the bone will change because the shape of the bone is changing. At any instant the bone will behave exactly like an elastic object, but the moving surfaces will cause the strain and stress to redistribute themselves slowly.

Model Predictions and Applications

The final model described above was initially proposed by Cowin and Van Buskirk (13) and applied to the surface remodeling induced by a medullary pin. Cowin and Firoozbakhsh (14) applied the model to the analytical prediction of the shape evolution of right hollow circular cylinders. The cylinders were idealized models of the diaphyseal region of a long bone. Hart et al. (12,15) formulated a computational model of the same situation using the finite element method. Cowin et al. (16) applied the theory to five situations in which the actual bone cross-section was modeled. These five actual bone cross-sections were taken from five animal experiments reported in the literature. Three of these animal experiments were modeled by changing a centric axial loading, and two were modeled by changing an eccentric axial load. Comparison of the results of the computational predictions with the animal experiments gave preliminary estimates of the remodeling rate coefficients occurring in the theory of surface remodeling.

REFERENCES

1. Menton, D.N., Simmons, D.J., Chang, S.L., Orr, B.Y. (1984). From bone lining cell of osteocyte - an SEM study. Anal Rec. 209:29
2. Chambers, T.J. (1985). The pathobiology of the osteoclast. J Clin Pathol 38:241.
3. Lanyon, L.E. (1984). Functional strain as a determinant for bone remodeling. (Cal Tiss Int 36:S56.
4. Gross, D. and Williams, W.S. (1982). Streaming potential and the electro-mechanical response of physiologically moist bone. J Biomech 15:277.

5. Salzstein, R.A., Pollack, S.R., (1987) Electromechanical potentials in cortical bone II. experimental analysis. J. Biomech 20:271.

6. El Haj, A.J., Skerry, T.M., Caterson, B., Lanyon, L.E. (1988). Proteoglycans in bone tissue: identification and possible function in strain related bone remodeling. Trans Orth Res Soc 13:538.

7. Ciba Foundation Symposium 124, (1986). "Functions of the Proteoglycans" Chichester, Wiley, p 5.

8. Yeh, C.K., Rodan, G.A. (1984). Tensile forces enhance prostaglandin E synthesis in osteoblastic cells grown on collagen ribbons. Cal Tiss Int 36:S67.

9. Nulend, J.D. (1987). "Cellular Responses of Skeletal Tissues to Mechanical Stimuli", (Doctoral Dissertation) Amsterdam, Free University Press.

10. Gross, S.B., Spindler, K.P., Brighton, C.T., Wassell, R.P. (1988) The proliferative and synthetic response of isolated bone cells to cyclical biaxial mechanical stress. Trans Orth Res Soc 13:262

11. Martin, R.B. (1972) The effects of geometric feedback in the development of osteoporosis, J Biomech 5:447.

12. Hart, R.T. (1983) "Quantitative Response of Bone to Mechanical Stress" Ph.D. Diss, Case Western Reserve Univ, Cleveland Ohio.

13. Cowin, S.C., Van Buskirk, W.C. (1979). Surface bone remodeling induced by a medullary pin. J Biomech 12:269.

14. Cowin, S.C., Firoozbakhsh, K. (1981). Bone remodeling of diaphyseal surfaces under constant load: theoretical predictions. J Biomech 14:471.

15. Hart, R.T., Davy, D.T., Heiple, K.G. (1984) A computational method for stress analysis of adaptive elastic materials with a view toward applications in strain-induced bone remodeling. J. Biomech Engng 106:342.

16. Cowin, S.C., Hart, R.T., Balser, J.R., Kohn, D.H. (1985) Functional adaptation in long bones: establishing in vivo values for surface remodeling rate coefficients. J. Biomech 12:269.

Tissue Engineering, pages 189–194
© 1988 Alan R. Liss, Inc.

MECHANISMS OF STRESS-INDUCED REMODELING

Harold Alexander, John L. Ricci,
Daniel A. Grande, and Norman C. Blumenthal

Department of Bioengineering
Hospital for Joint Diseases Orthopaedic Institute
New York, NY 10003

INTRODUCTION

It has long been noted that the musculoskeletal system responds to mechanical stimuli. Muscles, tendons, ligaments, and bone atrophy and lose mechanical strength and stiffness when they are either immobilized or underloaded for extended periods of time, as, for example, in cast immobilization, extended bedrest, or space travel. On the other hand, these same tissues hypertrophy and gain strength and mass in response to loading over extended periods of time—weightlifting, jogging, or bicycle riding, for example. However, if these same tissues are overloaded, tissue damage can result. We still cannot specify the kinds of stresses (shear, compression, tension), the magnitudes, and the frequency of application necessary to benefit these various tissues. There is, however, evidence that bone and cartilage respond to compression and that soft tissue responds to tension. Both compression and tension of the tissue cause shearing of the extracellular matrix (ECM). Data also indicate that, since bone and soft tissues are piezoelectric, the cell–mediating stimulus may be electrical in nature. This may also be cytoskeleton–related, since electrical stimulation has been shown to cause changes in the cytoskeleton.

Quantitative studies of bone remodeling have been performed both in vivo [1,2] and in vitro [3]. Carter et al. [4], Cowin and Firoozbakhsh [5], and Lanyon and Rubin [6] have all proposed theories relating mechanical stress to bone remodeling in a quantitative fashion. Carter et al. have proposed that shear stress (or strain energy, since the tissue is essentially incompressible) predisposes toward bone formation and that bone formation is inhibited or prevented by hydrostatic loading. Cowin and Firoozbakhsh have proposed that total strain governs bone remodeling. Lanyon and Rubin have proposed that static loading has little effect on bone remodeling, while dynamic loads causing strains in the functional dynamic strain range have dramatic effects. McLeod and Rubin [7] have proposed a strongly frequency-modulated effect of the application of electric fields. This may argue for a cyclotron resonance effect as proposed by Liboff [8]. None of these theories, however, is as yet generally accepted or can be used to predict the response to exercise or a particular surgical implant.

Quantitative studies of the response of fibrous tissue [9,10] and cartilage [11]

This work is supported by a grant from the G. Harold and Leila Y. Mathers Charitable Foundation and by National Institutes of Health grant NIH-NIAMSD R23 AR36544-03.

to mechanical stress have also been conducted. In general, however, the mechanical systems employed in the quantitative studies of stress-induced remodeling (SIR) of musculoskeletal tissue conducted so far have been poorly characterized. Only in an in vitro chondrocyte study did DeWitt et al. [12] make a serious attempt at characterizing the mechanics of their SIR system, but even their method was not well defined.

Most previous in vitro SIR studies have involved the use of explanted tissues or monolayer cultures. SIR of explants, because of interruption of blood supply, tissue microstructure, and the anisotropic nature of explant tissues, is unreliable, while SIR of monolayers involves cells cultured under conditions totally unlike the in vivo environment.

When any musculoskeletal tissue is exposed to mechanical stress, the result is deformation or strain of ECM. Our central hypothesis is that ECM strain, through attachments between the ECM, cell membrane, and cytoskeleton, induces changes in cytoskeleton configuration and that this causes the changes in levels of cyclic adenosine monophosphate (cAMP) that have been observed in SIR studies [13-15]. We have identified two mechanisms whereby changes in the cytoskeleton might be translated into changes in cAMP levels:

1. Enzymes associated with the adenylate cyclase/cAMP system have been observed to be associated with cytoskeleton elements. The relationship between cytoskeleton and cAMP, reviewed by Zor [16], represents one possible mechanism whereby cytoskeleton configuration changes result in changes in cAMP level.

2. Sundqvist and Ehrnst [17] found that the cytoskeleton is involved with movement of membrane–bound molecules, and Peters [18] and Zor [16] have postulated that the cytoskeleton may control cell sensitivity to hormones or growth factors. Changes in cytoskeleton configuration may thus translate into changes in receptor configuration, in turn causing changes in cell sensitivity to hormone stimulation and changes in cAMP levels.

In the study by Somjen et al. [14], physical strain was found to cause synthesis of prostaglandin E_2 by bone cells in culture, which resulted in increased cAMP production and increased ^3H–thymidine uptake. When cartilage cells were exposed to similar conditions, an increase in cAMP was observed that was independent of de novo prostaglandin synthesis. These findings suggest that either (1) there is a prostaglandin E_2–mediated pathway, exclusively for bone cells, that causes SIR, and that a different pathway exists in other musculoskeletal cells; or (2) the authors used strain levels or modes that are not appropriate to cause prostaglandin E_2 synthesis in cartilage cells. Since osteoblasts, fibroblasts, and chondrocytes normally exist in different ECMs and have different cell shapes and different cytoskeleton configurations, it is conceivable that they may respond to different levels or modes of physical strain. In vivo evidence suggests that these cells respond to compressive and tensile stresses in different ways.

In order to limit the extensive experimental variables involved with SIR studies, experiments must be conducted on the cellular level under controlled conditions. This can be accomplished using in vitro methods. Using recently developed techniques, it is possible to culture cells in reconstituted three-dimensional matrices that are similar to normal extracellular matrices and provide the cells with a more natural environment for these types of experiments.

We are currently conducting experiments which will investigate SIR, in vitro, on a cellular level. Special emphasis will be placed on the role of the intermediate filaments of the cytoskeleton as an integral part of a cellular mechanical transducer system.

CURRENT RESEARCH

Effects of Strain on the Cell Cytoskeleton

We are currently beginning a study in which fibroblasts, chondrocytes, and osteoblasts are cultured on collagen tapes, exposed to strains, and processed for immunohistochemical visualization of the intermediate filaments of the cytoskeleton. These cells will be compared to unstrained controls using morphometric analysis. Although this experiment involves the use of an unrealistic substitute for ECM, it will permit us to verify that strain of the cell substrate is transferred to the intermediate filaments of the cell cytoskeleton. We will also be able to determine the strain levels necessary for alteration of the cytoskeletons of the three different cell types involved and to determine the dynamics of cytoskeleton response to strain.

Effects of Strain on Connective Tissue Simulants

We are currently culturing fibroblasts, chondrocytes, and osteoblasts in sheets of reconstituted collagen ECM and exposing them to accurately controlled shearing and tension to determine the kinds of strain, strain magnitudes, and durations of application necessary to affect the growth and remodeling of connective tissues. These tissue simulants (cells in reconstituted ECM) have significant advantages over monolayer cultures in that (1) they can be molded and cut to shape, are isotropic, and have the necessary mechanical integrity to be cyclically strained; (2) they provide the cell with a three–dimensional matrix similar to normal ECM; and (3) cells produce their own ECM components within the reconstituted matrix, resulting in production of structures similar to normal tissues.

The cultures are being assayed for production of ECM, changes in cell growth rates, prostaglandin E_2 synthesis, cAMP levels, and other parameters. An additional series of studies will test the effects of epidermal growth factor (EGF) or calcitonin on mechanically strained cultures in an effort to determine whether mechanical strain changes cell sensitivity to growth factors.

Development of the System for Cyclically Straining Tissue Simulants

We have developed a mechanical system that allows us to expose tissue simulants to accurately controlled cyclic strains. The device consists of a set of chambers, with upper and lower reservoirs, which hold triplicate tissue simulants in a cell culture incubator. The tissue simulants are clamped over openings between the reservoirs using an O–ring seal-and-clamp system. A thin silicone rubber backing membrane, positioned under the tissue simulant, prevents pressure–induced fluid diffusion through the tissue simulant. Medium is cyclically pumped into the lower reservoir by a computer–controlled, stepping–motor–driven syringe pump. This mechanism sinusoidally inflates and deflates the silastic membrane and tissue simulant with controlled volumetric displacement. Pressure in the lower reservoir is monitored through a pressure sensor, and leaks are detected by sudden pressure drops.

To quantitatively determine the effect of tension and shear and combined quantities such as deviatoric strain energy on the remodeling response of the tissue simulants, it is necessary to perform the experiments in a number of different strain states. To obtain sufficient data to be useful, the strain state must be relatively uniform over a region containing a large number of cells. Using two flat membrane shapes, a circle and a "racetrack" configuration, controlled uniform strains are produced by inflation of a clamped specimen (see figure).

Figure.

Tissue simulant

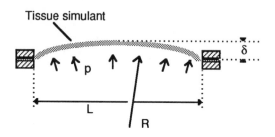

$$\sigma = f(p, R, t); \quad \varepsilon = \frac{S-L}{L}$$

$$S = 2R \sin^{-1}(L/2R)$$

$$R = \frac{L^2 + 4\delta^2}{8\delta}$$

Circular configuration

$$f(p, R, t) = \frac{pR}{2t}$$

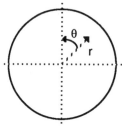

Racetrack configuration

$$f(p, R, t) = \frac{pR}{t}$$

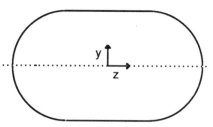

In center region:

$$\sigma_r = \sigma_\theta = \sigma$$

(equibiaxial strain)

Inflated shape is approximately a sphere for moderate inflation, $\varepsilon < 0.4$.

Away from semicircular ends:

$$\varepsilon_z = 0$$

(plane strain)

Inflated shape is a section of a cylinder.

Key	σ = stress	p = pressure
	ε = strain	R = membrane radius of curvature
	δ = displacement	t = tissue simulant thickness
		S = deformed length

The racetrack configuration (rectangular with semicircular ends) produces a state of plane strain in the region away from the ends [19]. The circular membrane inflation causes a state of equibiaxial tension in the central region at moderate strain levels [20]. For both configurations, it is possible, by controlling and monitoring the pressure and the membrane displacement, to determine the principle stresses, the principle strains, the maximum shearing stresses and strains, the total strain energy, and the distortional energy. Using the results from both experiment types at different stress and strain levels, it is possible to independently deduce the effects of tension, shear, strain energy, and distortional energy on cell function.

CONCLUSION

Stress-induced remodeling is a complex phenomenon which has been well demonstrated empirically but whose basic mechanisms have eluded researchers. It is our belief that the study of SIR must be taken to its most basic level, that of the cell and extracellular matrix. Using recently developed in vitro techniques, and the mechanical system described above, it is possible to produce musculoskeletal tissue simulants that closely approximate functional tissues and expose them to accurately controlled stress and strain conditions. This system will provide basic information concerning the cellular basis of SIR, enable us to interpret the results of in vivo studies, and allow us to test the hypotheses that arise from these experiments under controlled conditions.

REFERENCES

1. Lanyon, L. E., Goodship, A. E., Pye, C. J., and MacFie, J. H.,1982: Mechanically Adaptive Bone Remodelling. J. Biomechanics 15:141-145.
2. Goodship, A. E., Lanyon, L. E., and MacFie, J. H.,1979: Functional Adaptation of Bone to Increased Stress. J. Bone Jt. Surg. 61A:539- 546.
3. Steinberg, M. E., Gusenkell, G. L., Black, J., and Korostoff, E. 1974. Stress–Induced Potentials in Moist Bone In Vitro. J. Bone Jt. Surg. 56A:704-713.
4. Carter, D. R., Orr, T. E., Fyhrie, D. P:., and Schurman, D. J., 1987: Influences of Mechanical Stress on Prenatal and Postnatal Skeletal Development. Clin. Orth. Rel. Res. 219:237-250.
5. Cowin, S. C., and Firoozbakhsh, K., 1981: Bone Remodeling of Diaphyseal Surfaces under Constant Load: Theoretical Predictions. J. Biomech. 7:471-484.
6. Lanyon, L. E., and Rubin, C. T., 1984: Static vs. Dynamic Loads as an Influence on Bone Remodelling. J. Biomech. 17B(12):897-905.
7. McLeod, and Rubin, C. T., 1988: Stimulation of New Bone Formation by Low Frequency Sinusoidal Electric Fields. Trans. Ortho. Res. Soc., 34th Annual Meeting, Atlanta, p. 123.
8. Liboff, A. R., 1985: Geomagnetic Cyclotron Resonance in Living Cells. J. Biolog. Phys. 13:99-102
9. Amiel, D., Woo, S. Y., Harwood, F. L., and Akeson, W. H. 1982. The Effect of Immobilization on Collagen Turnover in Connective Tissue: A Biochemical–Biomechanical Correlation. Acta Orthop. Scand. 53:325-332.
10. Vailas, A. C., Tipton, C. M., Matthes, R. D., and Gart, M., 1981: Physical Activity and Its Influence on the Repair Process of Medial Collateral Ligaments. Conn. Tissue Res. 9:25-31.

11. Jones, I. L., Agreta, K., and Sandstrom, T., 1982: The Effect of Continuous Mechanical Pressure upon the Turnover of Articular Cartilage Proteoglycan In Vitro. Clin. Orthop. Rel. Res. 165:283-289.
12. De Witt, M. T., Handley, C. J., Oakes, B. W. and Lowther, D. A., 1984: In Vitro Response of Chondrocytes to Mechanical Loading. The Effect of Short Term Mechanical Tension. Conn. Tissue Res. 12:97-109.
13. Binderman, I., Somjen, D., and Shimshoni, Z., 1986: Growth Induction of Bone and Cartilage Cells by Physical Forces. In A. R. Hargens (ed.), Tissue Nutrition and Viability. Springer-Verlag, New York, pp. 121--134.
14. Somjen, D., Binderman, I., Berger, E., and Harell, A., 1980: Bone Remodeling Induced by Physical Stress Is Prostaglandin E_2 Mediated. Biochim. Biophys. Acta 627:91-100.
15. Binderman, I., Shimshoni, Z., and Somjen, D., 1984: Biochemical Pathways Involved in the Translation of Physical Stimulus into Biological Message. Calcif. Tiss. Int. 36:582-585.
16. Zor, U., 1983: Role of Cytoskeletal Organization in the Regulation of Adenylate Cyclic Adenosine Monophosphate by Hormones. Endocrine Rev. 4:1-21.
17. Sundqvist, K–G., and Ehrnst, A., 1976: Cytoskeletal Control of Surface Membrane Mobility. Nature 264:226-231.
18. Peters, R. A., 1956: Hormones and the Cytoskeleton. Nature 177:426.
19. Cook, T., Alexander, H., and Cohen, M., 1977: Experimental Method for Determining the 2–Dimensional Mechanical Properties of Living Human Skin. Med. Biol. Eng. Comput. 15:381-390.
20. Foster, H. O., 1966: Inflation of a Plane Circular Membrane. ASME paper no. 66–WA/RP–66.

Tissue Engineering, pages 195–200
© 1988 Alan R. Liss, Inc.

ELECTRICAL STIMULATION OF MUSCULOSKELETAL
GROWTH, REPAIR AND REMODELING

Jonathan Black, Carl T. Brighton and Solomon R. Pollack

Department of Orthopaedic Surgery, School of Medicine and
Department of Bioengineering, School of Engineering and
Applied Science, University of Pennsylvania,
Philadelphia, Pennsylvania 19104-6081

ABSTRACT: Observation of naturally occurring
electrical signals, of both matrix and cellular
origin, has led to broad studies of the role of
electrical phenomena in mediation of musculoskeletal tissue development and repair. Present clinical applications include faradic, capacitive and
inductive stimulation of acquired non-unions.
Laboratory results suggest that biological processes in many, if not all, tissues in the musculoskeletal system can be affected by such signals. Significant engineering problems remain, including determination of the mechanisms of production of naturally occurring signals, calculation
and measurement of local fields produced by externally applied signals and optimization of treatment for present and future clinical applications.

INTRODUCTION

A preeminent function of the musculoskeletal system in
animals and man is the provision of structural integrity
required for organ protection, shape maintenance and locomotion. In this role, the various tissues (skin, nerve,
muscle, tendon, ligament, cartilage and bone) must sustain
a variety of complex external and internal forces and react
in measured ways to protect and promote satisfactory function. Early observations, summarized in what has come to be
known as Wolff's Law, suggested a strong structural adaptive ability of bone, with structural remodeling processes

being mediated by applied stresses. This ability has been noted widely in musculoskeletal tissues. The emerging view is that the cell-directing physical signal, produced by structural transducers, has chemical, electrical and mechanical aspects; electrical effects may in fact be of primary importance.

ENDOGENOUS ELECTRICAL SIGNALS

In non-excitable tissues, there are two candidate electrical control signals:

Biopotentials:

All living tissues possess patterns of fixed potential gradients, typically 10^{-4} - 10^{-1} v/cm. Areas of higher cellular activity are negative with respect to surrounding "resting" tissue. Potentials are disturbed by trauma, becoming relatively more negative, and slowly return to normal as tissues heal and are remodeled. Cell death or uncoupling of oxidative phosphorylation produce rapid abolition of normal potential patterns while vessal ligation and/or local neurectomy leave them essentially unchanged. Their mechanistic origin is unclear.

Stress generated potentials:

Deformation of tissues produces transient polarizations, with voltage to strain ratios, dependent on deformation and deformation rate, reaching 1 - 2 v/ϵ. The initial polarization is such that areas in compression are relatively negative. All animal and plant tissues with long chain organic molecules (collagen, keratin, cellulose, etc.) display this effect. Production of these polarizations is fairly well understood: in dry tissue, classical piezoelectric phenomena may be present but streaming potentials dominate under physiological conditions.

EXOGENOUS STIMULATION MODALITIES

The prevalence of electrical phenomena in non-excitable tissues and a general correlation between negative polarization and increased cellular activity has led to a wide range of animal and clinical investigations of growth stimulation by applied electromagnetic signals. The numerous signals used may be grouped into three classes by

recognition of major field type and resultant current:
faradic (electric, net charge transfer (nct)), capacitive
(electric, no nct) and inductive (magnetic, no nct).

Each modality has been shown to have a broad range of
effects and dose-response relationships are known in many
biological models. However, there are important theoretical
and technical differences in their application. All three
modalities have been widely investigated in vitro and in
vivo, and have been successfully applied to clinical
problems.

Faradic stimulation:

Tissue elaboration occurs at a negatively charged
surface or implanted cathode. Imposed direct constant
currents of 1 - 20 μa are effective when cathodic potentials are below 1.5 v. Alternating and interrupted currents, with a cathodic bias, are effective but less so.
Cathode material and current density affect cell response.
The mechanism of effect is still unclear, probably involving electrodic changes (reduction in pO_2, increase in pH)
in the pericellular environment as well as electric field/
cell interactions.

Capacitive stimulation:

Externally applied planar electrodes with symmetric or
assymmetric alternating electric currents at frequencies
above 10 kHz can induce local displacement currents with
associated electrical fields between 10^{-3} and 1 v/cm,
dependent upon tissue type. These fields can produce direct
polarization effects on extracellular molecules (enzymes,
substrates, etc.) as well as altering structure and function of membrane bound receptors and ion channels; however,
membrane impedance excludes them from cell interiors. A
variety of metabolic and growth acceleration effects are
seen in vitro and in traumatized animal tissue models.

Inductive stimulation:

Time varying magnetic fields, applied by Helmholtz
coils, with or without magnetic susceptors, can produce
internal electrical fields of the same strength as capacitive stimulation. However, since tissue has the magnetic
permeability of air, lower frequencies can be used. Most
signals are pulsed, with carrier frequencies below 5 kHz.

The range of biological effects observed is similar to that
produced by capacitive stimulation but the mode of signal
effect is unclear: it may be directly electrical (including
intracellular dB/dt effects) or by paramagnetic or cyclotron coupling (in the presence of the earth's magnetic
field) or by some combination of effects.

STIMULATION IN CELLULAR AND ANIMAL MODELS

Bone has been the primary tissue of interest, although
more recently attention has been turning to soft tissues.
While all three modalities show repeatable dose-response
effects, often in similar models, there is, as yet, no
unifying theory of effect. Changes in cell habit, replication and synthetic rates have been observed in vitro,
frequently associated with changes in Ca^{++} ion management.
In vivo models show a variety of changes in tissue elaboration and calcification; in fracture healing models the
effects are only seen in early and mid-phases. Early animal
studies show positive effects in prevention and reversal of
osteo- porosis and indications of beneficial effects on
healing of articular cartilage and peripheral nerve injury.
These results have led to a wide range of clinical studies.

CLINICAL APPLICATIONS

Devices have been approved by the US Food and Drug
Administration for routine use in a limited number of
clinical indications. These indications (and stimulation
modalities) are: acquired non-union (faradic, capacitive
and inductive), congenital non-union (inductive), and
spinal fusion (faradic). Clinical trials of treatment of
delayed union, fractures at risk (high energy, comminuted),
congenital non-union, avascular femoral head necrosis and
post-menopausal spinal osteoporosis are under way with at
least one treatment modality each. However, significant
problems remain, many of which require extensive engineering analyses.

CLINICAL PROSPECTS

The close association between electrical phenomena in
nonexcited tissue and growth, repair and remodeling processes has led to two hypotheses:
1. Naturally occurring growth processes depend upon
the local electrical environment within tissues.

2. Manipulation of this environment can beneficially affect cellular behavior.

To the degree that these hypotheses prove to be true and the resulting theses can be reduced to engineering practice, we can expect to see applications of electrical stimulation in treatment of musculoskeletal disability. In the near term, some form of electrical stimulation may come to be standard adjunctive therapy in all modes of fracture treatment; in the longer term it is expected that electrical stimulation will come to play a major role in primary treatment of a wide range of congenital and acquired musculoskeletal disabilities.

TISSUE ENGINEERING OPPORTUNITIES

In the sense that "tissue engineering" means controlling synthesis of living tissue by physical means, all of the field of electrical stimulation of growth, repair and remodeling is a part of Tissue Engineering. Questions requiring engineering analysis run the gamut from fundamental (mechanisms of field/molecule/cell/tissue interactions) through applied (magnitudes of local fields and current densities) to clinical (design of optimized clinical treatment signals and devices). Furthermore, studies of electrical stimulation of tissue activity offer valuable opportunities to weigh the relative contributions of chemical, mechanical and electrical stimuli and their interactions. Progress to date has been marked by active collaboration between engineers, biological scientists and clinicians; future achievements depend upon continued activities of strong interdisciplinary research groups.

REFERENCES

It is not possible in this small space to summarize more than thirty years of modern era research on electrical stimulation of growth, repair and remodeling. Thus, only the following general references are given. Detailed information on points in this review may be found within them and their bibliographies:

Becker, RO, Marino AA (1982). "Electromagnetism and Life." Albany: State University of New York
Black, J (1987). "Electrical Stimulation: Its Role in Growth, Repair and Remodeling of the Musculoskeletal System." New York: Praeger.
Brighton CT (ed) (1984). "Symposium on Electrically Induced Osteogenesis." Orthop Clin NA 15(1):1
Brighton CT, Black J, Pollack SR (eds) (1979). "Electrical Properties of Bone and Cartilage. Experimental Effects and Clinical Applications." New York: Grune & Stratton.
Chiabrera A, Nicolini C, Schwan HP (eds) (1985). "Interactions Between Electromagnetic Fields and Cells." New York: Plenum.
Okada K (ed) (1984). "Piezoelectricity of Bone and Electrical Callus. Experimental and Clinical Studies by Professor Iwao Yasuda." Sapporo: Fuji.

Tissue Engineering, pages 201–206
© 1988 Alan R. Liss, Inc.

BIOMATERIAL-TO-BONE INTERFACES

J. E. Lemons

Department of Biomaterials
University of Alabama at Birmingham
Birmingham, AL 35294

ABSTRACT   Surgical implants fabricated from titanium
and alloy, carbon and carbon-silicon, and aluminum oxide
and calcium phosphate ceramics under controlled conditions
and precisely placed in laboratory animals have shown direct
biomaterial-to-bone interfaces when evaluated by optical and
electron microscopy. This interfacial tissue development and
stability has been shown to depend upon both the chemistry
of the biomaterial surfaces and the specific details of
functional force transfer. Attempts to definitively separate
the chemical-biochemical from the mechanical-biomechanical
property relationships demonstrate a complex synergism. In
this regard, very dilute concentrations of organic and
inorganic impurities alter tissue reactions and directly
influence the tissue interface development. Also, it is known
that bone can be changed to fibrous tissue through
biomechanical pathways. Interfacial motion, especially if the
motion is immediately post-placement, can result in similar
interfacial tissues, i.e., similar to impurities and improper
stress profiles. Comparisons of controlled laboratory animal
and somewhat uncontrolled human retrieval specimens have
shown some very interesting similarities and differences that
can be associated with biomaterial and clinical practice
characteristics. These observations which have resulted in a
basic working hypothesis will be presented in this paper.

INTRODUCTION

Surgical implant device retrievals provide opportunities to
observe biological tissue responses from controlled conditions
within laboratory animals and somewhat uncontrolled implant

conditions in humans that have previously received reconstructive procedures.(1-9) The laboratory animal evaluations include time-based sequences where tissue development and maturation can be correlated with interfacial breakdown from bone tissue to fibrous tissue as a function of the implant biomaterial and the device implant conditions. In contrast, human devices that are removed only irregularly, sometimes provide situations where the device was functional and not significantly compromised by clinical conditions (e.g., infection) at the time of removal. Dental implant tooth root replacements are one example where devices have been removed with the directly associated tissues intact because of compromised intraoral superstructure failures rather than implant conditions. Dental bridge-to-implant-connector fractures that necessitate implant removal have provided some device and tissue samples for interfacial evaluations. These include a wide range of biomaterials and designs and thereby the opportunity to compare similar biomaterials and designs from both laboratory animal and human retrievals. To date, more than 250 dental and 1300 orthopaedic devices from humans have been evaluated.

Implant interface evaluations, over the years, have shown mixtures of fibrous tissue and bone depending upon a wide range of variables. Only recently, within the last two decades, have implant surfaces and clinical procedures come under careful scrutiny. In part, this has been because of the interest in developing direct bone interfaces and improved force transfer conditions, i.e., osteointegration of functional implant devices.

Attempts to evaluate the detailed tissue characteristics of biomaterials in bone have been complicated by the complex multifactorial basis for tissue development and destruction. For example, how does one isolate a bone implant from mechanical stress in a vital functional bone? Also, is it possible to place chemically clean implants under normal operating room conditions? In both cases, the answer is, "Almost".(10) Laboratory animals are, of course, the best opportunity.

Dental implants have undergone two very significant changes during recent years. The more common one-stage implant device that was loaded shortly after implant placement has been partially replaced by two-stage systems. Two-stage devices are placed and allowed to heal without interconnection to the oral environment during stage-one. Stage-two device attachment is made after an extended period at three to six months. Also, the importance of maintaining very clean surgical conditions has now been emphasized by most clinicians.

In contrast to the dental two-stage systems, orthopaedic

devices function, although restricted, shortly after placement, e.g., total hips and knees.

Because of the availability of many different devices and implant retrievals, a working hypothesis was developed. Selected biomaterials such as the titanium, carbon, and aluminum and calcium phosphate based ceramics would form osteointegrated interfaces under conditions of a properly prepared clean surface and restricted functional loading. A second hypothesis was that this interface could be altered to become fibrous tissue because of improper mechanical stress and/or implant surface impurities from the time of surgery or from biodegradation phenomena. The latter conditions have been the focus of recent studies and will be emphasized in this paper.

METHODS

Laboratory animal implant forms have included particulates, porous surfaces, discs, rods, and functional devices fabricated from commercial purity titanium (CP Ti), titanium-6 aluminum - 4 vanadium (ELI Grade), cobalt (F-75) and iron base alloys (316L) high purity pyrolytic carbon and carbon-silicon compound, and high purity aluminum oxide (alumina and sapphire) and calcium-phosphate (hydroxylapatite and tricalcium phosphate) ceramics. These were maintained under clean and sterile conditions and implanted under controlled laboratory animal conditions. Surgical protocols were set to avoid implant contamination. In some cases the surfaces were contaminated with known amounts of inorganic impurities or forced to biodegrade in vivo.(11,12) Surface conditions of the implants were investigated by optical and electron spectroscopy methods to confirm clean or contaminated conditions.(13) Metals and alloys were passivated following ASTM methods, carbons were radiation sterilized, and ceramics were not exposed to moisture and were sterilized using dry heat environments.

Human implant devices were processed under the standard dental oral surgical or medical orthopaedic surgical environments. Metals and alloys were provided sterile by the manufacturer as were the carbons. Ceramics were dry heat sterilized in most situations, although this was not always known.(14)

The dental implant designs included endosseous blades of titanium (Oratronics, Park, and Customs) and titanium-aluminum-vanadium (MITER) and root forms of aluminum oxide, carbon, titanium and alloy, and calcium phosphate ceramics (MITER, Core-Vent, Kyocera, Omnii, Biotes, Vitredent and Carbomedics, Neodontics, Integral and Calcitek-coated) devices.

RESULTS AND DISCUSSION

The general observations associated with the biomaterial-to-tissue interfaces from laboratory animal and human implant device retrievals are summarized in Table 1.

TABLE 1
SUMMARY OF BIOMATERIAL TO TISSUE
INTERFACES UNDER RETRIEVAL CONDITIONS

Observations and Biomaterials							
Implant Type	Ti	Ti-6A1-4V	C	C-Si	Al$_2$O$_3$	Ha	TCP
Laboratory Animal	B	B	B	B	B	B	B/B/FT
Human Implant	B,FT,B/FT	B/FT	FT	FT	B/FT	B/FT	FT/B

B = Bone, FT = Fibrous Tissue. When B is listed alone, this indicates more than 70 percent of interface showing bone contact. These do not include contaminated interfaces, and only represent the conditions at the time of retrieval.

In general, these various biomaterials provide three different types of surface conditions in contact with the bone. The titanium and alloy have a relatively thin titanium oxide surface (Ti_xO_y) over an electrically conductive substrate. The carbon and carbon-silicon are also electrically conductive but have a quite different surface chemistry compared to the metallic oxides. These surfaces are carbon or a compound of carbon and silicon. Aluminum oxide and calcium phosphate ceramics are not electrical conductors. The aluminum oxides would be classed as inert (Al_2O_3) while the calcium phosphates are surface active (bioactive). Another way to consider these interfacial characteristics is to compare the biomaterial surface and the tissue. This is done in Table 2.

Interestingly, all of these biomaterials have shown "osteointegration" within laboratory animal models. One might ask then, why have the carbon clinical implants shown fibrous tissue interfaces? Also, why was this the case for most of the titanium endosseous blades? In the author's opinion, the carbon devices, were inadvertly contaminated during sterilization and placement procedures. This predisposed the carbon implants to the evolution of a fibrous tissue interface because of chemical-biochemical reactions. In contrast, most titanium endosseous blades were mechanically loaded shortly after placement (1 to 6 weeks) inducing motion at the implant-to-tissue interface.

In both cases however, where some these same implants were placed under clean conditions or were not inappropriately loaded, bone interfaces were present. Interestingly when properly restored, the fibrous tissue interfaces did not predispose the endosseous blades to clinical failures.

TABLE 2
BIOMATERIAL SURFACES AND TISSUE
RESPONSES

Biomaterial	Surface	Activity
Ti	Ti_xO_y	A/P
Ti-6Al-4V	Ti_xO_y	A/P
C	C	P
C-Si	C-Si	P
Al_2O_3	Al_2O_3	P
HA	$Ca_{10}(PO4)_6(OH)_2$	A
TCP	$Ca_3(PO4)_2$	A

A = Active (possible interfacial bonding)
P = Passive (inert biomaterial)

Laboratory animal studies have shown that bone tissue responses can depend upon conditions of mechanical motion, unacceptable stress, and impurity or implant biodegradation products. It now appears that the early requests from the research community to maintain chemically and mechanically clean implants, were much more important than originally recognized.

In this area, it is apparent that three types of quantitative data need to be developed under prospective protocols: (1) quantitation of interfacial micromotion and the limits that can be tolerated during bone healing, modeling, and remodeling; (2) biomechanical limits of interfacial stress states along passive-inert and bioactive-bonded implant interfaces; and (3) the dose-response-time relationships for controlled impurities and biodegradation products for the various classes of biomaterials. These types of studies are strongly recommended.

REFERENCES

1. McKinney RV Jr., Lemons JE (1985). "The Dental Implant." Littleton: PSG Publishing.
2. Branemark PI, Zarb GA , Albrektsson T (1985). "Tissue Integrated Prostheses." Chicago: Quintessence Books.
3. Koth DL and McKinney RV Jr. (1981). The single crystal sapphire endosteal dental implant. In Clark J W (ed.): "Clinical Dentistry," New York: Harper and Row, Chap 53.
4. Lemons JE, Niemann KMW, Weiss AB (1976). Biocompatibility studies on surgical grade titanium, cobalt and iron base alloys. J Biomed Mater Res 10:549.
5. Meffert RM, Block MS, Kent JN (1987). What is osseointegration? Int J Perio and Rest Dent 4:9.
6. Smith DC, Williams DF (1982). "Biocompatibility of Dental Materials IV." Boca Raton: CRC Press.
7. deGroot K (1983). "Bioceramics of Calcium Phosphate." Boca Raton: CRC Press.
8. Dennissen H, Mangano C, Venini G (1985). "Hydroxylapatite Implants." Padua, Italy: Piccin.
9. Vincenzini P (1983). "Ceramics in Surgery." Amsterdam: Elsevier.
10. Huskes R, Roberts VL (1987). Anniversary issue on bone biomechanics. J of Biomech: p 1015.
11. Lucas LC, Bearden LF, Lemons J E (1985). Ultrastructural examinations of in vitro and in vivo cells exposed to solutions of 316L stainless steel. In Fraker A, Griffin C (eds): "ASTM STP 859," Philadelphia: p 208.
12. Lemons JE, Niemann KMW, Weiss AB (1976). Biocompatibility studies on surgical grade Ti, Co and Fe base alloys. J Biomed Mater Res 10: 549.
13. Macon ND, Lucas LC, Lemons JE, Henson PG (1986). In vivo tissue response to Ti-6Al-4V/Co-Cr-Mo implants. Trans Soc for Biomat 9:50.
14. Lemons JE, Chamoun E (1985). Evaluations of retrieved dental implant devices. In B.W. Sauer (ed): "Biomedical Engineering IV." New York: Pergamon Press, p 145.

Tissue Engineering, pages 207–208
© 1988 Alan R. Liss, Inc.

Summary/Discussion

Bone

Jonathan Black

In addition to contributing to design and fabrication of tissue and hybrid implants, engineers are turning their attention towards analysis of normal and pathological mechanisms of growth, repair and remodeling.

Adaptive remodeling, particularly of hard tissue, has attracted considerable attention since Julius Wolff first tried to codify a "law" linking function to structure. Variously stated, Wolff's Law essentially notes that structure, once created by genetics and developmental stimuli, alters in response to applied loads.

Two engineers' views of the closed-loop control system are represented here. Dennis Carter's approach depends upon local stress intensity functions. Using a variety of fitting functions and the iterative processes of modern finite element analysis computer programs, he is able to show how bony anatomy in the proximal femur and in healing long bone fractures can be generated from simple initial conditions. Of particular note is his association of cartilage with regions of compressive (hydrostatic) stress and bone with regions of shear (deviatoric) stress. In the view of many, including this reviewer, his position is weakened by overdependence upon parametric fitting processes, by the use of stress as a necessary and sufficient stimulus, and by the near exclusion of biological (endocrine, apocrine, etc.) control processes.

By comparison, Stephen Cowin's theory is modular in form. Emphasizing total strain history of mineralized tissue, it depends upon a balance between two activity functions — one for osteoblastic accretion and the other for osteoclastic resorption of bone. These functions act at surfaces, in a manner consistent with bone biology, and depend, in a generalized relationship, upon strain history as well as upon chemical, electrical and biochemical stimuli. A major victory of their theory lies in its excellent correlative explanation of Lanyon's experimental observations of orvine radial remodeling after segmental ulnar resection.

In the final analysis, both approaches are overly reliant on mechanical stimuli. This reflects, most probably, the early disciplinary training of the two speakers. Mechanical engineers and their orthopedic colleagues tend to emphasize mechanical stimuli of biological processes, cell biologists focus on chemical/biochemical factors while electrical engineers and electrobiologists emphasize electrical phenomena.

Even in a simple situation, such as the local host response of mineralized tissue to a synthetic implant (Figure 1), one may identify four spectra of physical induction factors: chemical, electrical and mechanical, the latter including both stress and relative motion. One may individually correlate biological observations with relative values along each of these spectra. A central challenge in tissue engineering is to deduce each of the appropriate functional relationships and to provide an overall statistical constitutive equation linking all stimuli to the biological responses.

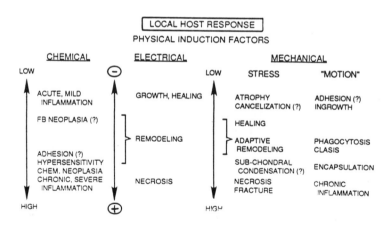

Figure 1. Physical induction factors in local host response. Reproduced by permission of Churchill-Livingstone, Inc. From Black, J.: Orthopaedic Biomaterials in Research and Clinical Practice (1988).

Tissue Engineering, page 209
© 1988 Alan R. Liss, Inc.

V. Transplants and Artificial Organs

Introduction. . . . . George Boder

Tissue engineering is a multi-faceted problem involving a number of
scientific disciplines. This meeting underscores the importance of multidisciplinary approaches in a new and rapidly expanding area that can provide
solutions to solve practical problems. The areas treated included aspects of
endothelial-smooth muscle cell interaction, and approaches to gain an
understanding of the complex interactions of cells with one another in
changing environments.

Transplantation of cells or use of cells to form tissues for subsequent use as
organs or prosthetic devices present fundamental problems: What is the
state of differentiation of the cells? How will the state of differentiation be
affected at any given implantation site? Will the site of implantation
provide an environment that maintains the implanted tissue in an adequate
equilibrium leading to either senescence or hyperplasia? Specific examples
were addressed, e.g. endothelial cells placed on the wrong form of collagen
have altered phenotypic expressionand function. In spite of this, reasonable
therapeutic targets have been established, such as vascular grafts and nerve
cell regeneration devices (bridges).

R.A. Peura described problems encountered with transplantation of islet
cells, such as what could be done with excessive secretion. Various forms
of islet encapsulation and immobilization were reviewed (hollow fibers,
gels, etc.). These have been purported to avoid immune rejection and
overgrowth with fibroblasts. These systems have been the subject of
intense research for a number of years, but progress to a stage equated with
utility has not been made. Some optimism was generated during discussions about new biocompatible polymers.

The benefits of insulin pumps are limited by the lack of sensitive glucose
sensors for in-situ monitoring, a problem that has been evident for a number of years. This is a basic tissue engineering problem and a worthwhile
goal for continuing efforts.

Tissue Engineering, pages 211–216
© 1988 Alan R. Liss, Inc.

BIOARTIFICIAL ORGANS

P. M. Galletti and P. Aebischer
Artificial Organs Laboratory, Brown University
Providence, Rhode Island 02912

ABSTRACT   Bioartificial organs are hybrids between
prosthetic devices and natural organs, combining the
distinctive features of the two major approaches to the
replacement of lost body function: artificial organs
and live tissue transplantation. Bioartificial organs
can be implants of live cells, surrounded by a permselective synthetic membrane which isolates the
transplant from its environment. In that case, the
polymer capsule allows the placement of living tissues
in an immuno-incompatible recipient without the need
for pharmacologic immunosuppression. The functional
purpose of the device is humoral interaction between
the graft and the host recipient. Therefore the
membrane material surrounding the transplant must be
permeable to appropriate solutes (nutrients, messengers
and hormones) so that the encapsulated tissue may live
and function indefinitely in a sequestered environment. At the same time, the membrane must prevent the
entry of immunoglobulins and lytic factors of the
complement and preclude the escape of immunogenic
molecules, as well as the inward migration of
leucocytes and phagocytes. The term "bioartificial
organ" is also used to designate the endpoint of organ
reconstruction on polymer scaffolds. Organ
reconstruction can be initiated in vitro when cell
seeding or proliferation requires tightly controlled
culture conditions, e.g. alternance of temperature to
eliminate a contaminant cell population. It can be
induced in vivo when the morphological, chemical and
mechanical characteristics of the biomaterials used are
appropriately tuned to the development of endogenous
cell populations.

The general definition of bioartificial organs covers
all replacement parts which purposely combine living cells
or tissues with synthetic biomaterials in the replacement of

a body part or function. The concept of bioartificial
organs – or hybrid artificial organs as they were called
initially – dates back to the mid-1970's (1,2), although
precursors can be identified at a much earlier date (3).
From a descriptive viewpoint, we can distinguish between
"structural devices" and "interactive devices." In
structural devices, the artificial, engineered component
provides the initial physiologic function, while the live
cells and the tissues they form serve to stabilize,
passivate or eventually replace the synthetic biomaterial.
In interactive devices, the physiologic functions of the
implant are performed by its living components, whereas the
synthetic membrane serves merely to establish an
immunological barrier between the transplant and its host
while still allowing solute exchanges.

BIOARTIFICIAL ORGANS

Structural Devices	Interactive Devices
Living cells serve to anchor or passivate the synthetic material	Living cells provide the major physiological functions of the implant
Cell attachment and ingrowth characterize the integration of the synthetic material within the host	Immunoisolation by synthetic membrane must preserve solute exchange with the environment
Continuity host-transplant	Discontinuity host-transplant
The ultimate goal is <u>controlled</u> external tissue reaction	The ultimate goal is <u>minimal</u> tissue reaction

Endocrine tissues initially provided the most
attractive challenge for bioartificial organs since such
implants involved only a small mass of tissue, and secreted
an easily identifiable substance. There were also well-
established animal models of endocrine deficiency in which
the pharmacological effectiveness of a membrane-sequestered
transplant could be unequivocally ascertained. Nowadays one

can identify a wide range of applications of the
bioartificial organ concept of over a scale of growing
metabolic complexity. In the simplest form of device, the
cell-biopolymer composite provides primarily the passive
function of a mechanical spare part. The metabolic activity
is largely restricted to providing for the nutritional needs
of the living components of the device, e.g. the ciliated
epithelium cells cultured on a tracheal prosthesis.
Bioartificial local release devices include cells for their
ability to synthetize a specific molecule which is needed in
a defined location, e.g. the neurotransmitter dopamine near
the corpus striatum, or enkephalins in the pain modulating
regions of the central nervous system. Secreting tissue
implants in general release devices serve the same purpose
as open-loop drug delivery systems, except that they are
limited to molecules synthetized by natural or genetically
engineering cells which do not require chemical feedback for
secretion control. Bioartificial hormonal devices are those
where the release of a specific product is regulated by
humoral messengers from the external environment. The
interplay of signal and response insures the homeostasis of
a particular solute, e.g. blood glucose regulation by
insulin release from a bioartifical pancreas. The most
complex devices, which are still at the level of early
prototypes, will be implantable biochemical reactors
capable of processing simultaneously multiple chemicals or
metabolites. This is the challenge of the bioartifical
kidney and the bioartificial liver now, the bioartificial
lung and bioartificial placenta at some later time.

Two fabrication techniques currently dominate the
technology for surrounding live tissue with a synthetic
membrane of appropriate permselectivity: microencapsulation
and macroencapsulation. In microencapsulation, a polymer
gel is built around each cell or cell cluster. The resulting
suspension of tissue fragments is then injected in the
peritoneal cavity or the blood vessel which feeds the target
site for the transplant. Macroencapsulation starts with a
preformed tubular membrane closed at one end, then filled
with donor tissue fragments or isolated cells, capped
tightly at the other end and surgically implanted at the
appropriate site in the body.

Both systems allow the exchange of small nutrient
molecules while at the same time preventing immune
recognition by the recipient of the xenograft. Microencapsulation offers the advantage of a thin membrane
surrounding a small mass of tissue, thereby reducing to a
minimum the diffusion distances for solutes moving in and

out of the transplant. It has been claimed to be superior
to macroencapsulation in preserving initial tissue viability
and secretory response time. The technique requires complex
in vitro processing if the individual cells to be
transplanted must be dissociated from a solid organ. It
cannot easily be adapted to irregular tissue fragments, and
there remain questions about the long-term stability and
bioacceptance of the synthetic polymers used to insulate the
individual implants.

Macroencapsulation relies on materials of well-
established inertness, durability and compatibility in
various body locations, and is mechanically adaptable to a
wide variety of tissue structures. It requires special care
in handling to insure high cell density and reliable closure
of the polymer envelope. The wall thickness and the inner
diameter of the tube are critical determinants of diffusion
distances and must be adapted to the metabolic needs of each
tissue. Some macroporous, textured membranes allow the
ingrowth of a capillary network which brings external
metabolic support within a few microns of the implant.

The concept of bioartificial organs has been validated
experimentally with secretory cells from the pancreas (4,5),
thyroid (6), parathyroid (6,7,8), thymus (9), adrenal (6),
pituitary (10), mid brain tissue (11), renal tubule (12) and
liver parenchyma (13,14,15). There is no a priori reason
why it could not be applied to any endocrine tissue (16).

The dominant issue in the design of bioartificial
organs is that of membrane material biocompatibility, i.e.
the ability of the synthetic material to support cell growth
and stablization within the capsule, while at the same time
eliciting only a minimal, rapidly stabilised inflammatory
reaction in the tissues surrounding the polymer-protected
transplant. The criterion for biocompatibility is the
ability of the material to lead to permanent acceptance of
the xenograft while preserving its functional capability.
This is not easy to achieve, considering the cumulative
complexity of the process. Failure can originate at any
step in the sequence of tissue procurement from a donor,
dissociation into individual cells or cell clusters, device
fabrication under sterile conditions, induction of an animal
model of disease, microsurgical implantation procedure, early in vivo survival of the grafted tissue and late host reaction to the implant. If one examines the literature
carefully, one will conclude that at the present stage of
technology, a failure rate up to 50 percent is not
uncommon. Such an outcome is not a major obstacle for
laboratory research, but it is not acceptable for a clinical

procedure. Hence further refinements are needed before bioartificial organs assume a therapeutic value.

From the viewpoint of tissue engineering, the two major problems are: the reorganization of the dissociated cells or tissue fragments within the capsule into a functional tissue interfacing with the inner skin of the membrane; and the need for a self-limiting reactive tissue organization around the implant which does not choke off nutritional support for the graft and bidirectional solute exchange with the environment. For both processes, the critical elements are the structure of the wall, and the surface and permeability properties of the membrane material.

In the history of artificial organ design and fabrication, there is a standard path of evolution from off-the-shelf materials selected because of availability and convenience to specialty materials structured for a specific biological goal. For bioartificial organs we have hardly entered the second phase yet. Therefore we expect a period of engineering research focussed on the molecular configuration of candidate polymer surfaces and the microarchitectonics of their surface and wall structure. This effort needs to be coupled with biological research on cell surface receptors, identification of attachment factors, preservation of cell polarity, differentiation, and functional linkages with adjacent cells. This is a daunting challenge for the tissue engineers of the next decade, one which will require a continuum from materials science to surface chemistry, biomechanics, cell biology, organ physiology and microsurgery. The facilities and research equipment needed cover an equally broad range, suggesting that the entire continuum may not be provided by a single group or institution. The commercialization of products including a live component is also fraught with complex uncertainties. Yet all of these issues must be resolved before the turn of the century. Thus it is quite timely that they be addressed in an integrated fashion under a national program.

REFERENCES

1. Chick WL, Like AA, Lauris V, Galletti PM, Richardson PD, Panol G. Mix TW, Colton CK (1975). A hybrid artificial pancreas. Trans Amer Soc Artif Int Organs 21:8.
2. Tze WJ, Wang FC, Chen LM, O'Young S (1976). Implantable artificial endocrine pancreas unit used to restore normoglycaemia in the diabetic rat. Nature 264:466.
3. Bisceglie V (1933). Über die antineoplastische Immunitat. II. Mitteilung. Über die Wachstumsfähigkeit der

heterologen Geschwülste in erwachsenen Tieren nach
Einpflanzung in Kollodiumsäckchen. Ztschr f. Krebs-
forschung 40:141.

4. Altman JJ, Houlbert D, Callard P, McMillan P, Solomon
   BA, Rosen J, Galletti PM (1986). Long term plasma glucose normalization in experimental diabetic rats using
   macroencapsulated implants of benign human
   insulinomas. Diabetes 35:625.

5. O'Shea GM, Sun AM (1986). Encapsulation of rat islets
   of Langerhans prolongs xenograft survival in diabetic
   mice. Diabetes 35:943.

6. White DC, Trepman E, Kolobow T, Bowman RL (1979). The
   blood-membrane bioartificial system: Hormone production
   by human endocrine adenomas perfused with sheep blood.
   Trans Am Soc Artif Int Organs 25:32.

7. Aebischer P, Christenson L, Russell PC, Panol G, Monchik
   JM, Galletti PM (1986). A bioartificial parathyroid.
   Trans Amer Soc Artif Int Organs 32:134.

8. Darquy S, Sun AM (1987). Microencapsulation of parathyroid cells as a bioartificial parathyroid. Trans Am
   Soc Artif Intern Organs 33:356.

9. Christenson L, Aebischer P, Galletti PM (1987). A bioartificial thymus as a potential immunomodulator. Artif
   Organs 11(4):338.

10 Hymer WC, Wilbur DL. Page R, Hibbard E, Kelsey RC, Hat-
   field JM (1981). Pituitary hollow fiber units in vivo
   and in vitro. Neuroendocrinology 32:339.

11. Aebischer P, Winn SR, Galletti PM (1988). Transplanta-
    tion of neural tissue in polymer capsules. Brain Re-
    search, in press

12. Aebischer P, Ip TK, Miracoli L, Galletti PM (1987).
    Renal epithelial cells grown on semi-permeable hollow
    fibers as a potential ultrafiltrate processor. Trans
    Amer Soc Artif Int Organs 33:96.

13. Wolff CFW (1980). Cells cultured on artificial capil-
    laries and their use as a liver assist device. Artif
    Organs 4:279

14. Hager JC, Carman R, Porter L. Stoller R, Leduc E, Gal-
    letti PM, Calabresi P (1983). Neonatal hepatocyte cul-
    ture on artificial capillaries: A model for drug meta-
    bolism and the artificial liver. asaio J 6:26.

15. Sun AM, Cai Z, Shi Z, Ma F, O'Shea GM, Gharapetian H
    (1986). Microencapsulated hepatocytes as a bioarti-
    ficial liver. Trans Am Soc Artif Int Organs 32:39.

16. Sun AM (1987). Encapsulated versus modified endocrine
    cells for organ replacement. Trans Am Soc Artif Int
    Organs 33:787.

Tissue Engineering, pages 217–222

O_2 EFFECTS ON PANCREATIC ISLET INSULIN SECRETION IN HYBRID ARTIFICIAL PANCREAS

Clark K. Colton, Keith E. Dionne,
and Martin L. Yarmush

Department of Chemical Engineeirng
Massachusetts Institute of Technology
Cambridge, MA 02139

The complications of diabetes are thought to arise from poor control of blood glucose concentration because of inadequate pancreatic insulin secretion, motivating novel therapeutic approaches to improve insulin delivery as compared to daily insulin injections. These approches fall into two categories: (1) pure artificial systems, and (2) hybrid devices which incorporate living pancreatic tissue. Artificial systems include external and implantable pumps and sustained release polymeric systems which provide a constant or variable insulin infusion rate. Development of an implantable glucose sensor may permit a closed-loop feedback system which responds to changes in blood glucose concentration. The hybrid systems make use of immunoisolation to circumvent immunological rejection problems of transplantation by interposing a semipermeable membrane, which passes glucose and insulin but retains antibodies and lymphocytes, between transplated pancreatic islet or β-cells and the host tissue [1]. In principle, the use of living islets offers the advantage of physiologic feedback regulation of insulin secretion in response to changes in glucose and other secretagogue concentrations in the blood.

A wide variety of bioengineering problems must be solved before hybrid devices can reach the point of clinical application. These include (1) development of large-scale processes for islet/cell procurement from suitable xenogenic sources, (2) maintenance of cell viability and insulin secretion capability during isolation and thereafter within the device for long periods of time, (3) preparation of membrane materials which eliminate bioincompatibility phenomena associated with host-device interactions, (4) fabrication of membranes with suitable geometries and physical properties, and (5) design of devices which provide satisfactory insulin secretory performance with feedback control over long periods of time.

Several design variations have been examined for use as pancreatic replacements. One variant is extravascular implantation in which the islets are sandwiched between two synthetic membrane sheets, placed in the lumen of hollow fiber membranes, or microencapsulated in a hydrophilic gel or capsule. Implantation sites include subcutaneous, intra-peritoneal, and intrasplenic locations. The major problem is fibroblastic overgrowth of the device to form an avascu-lar capsule which surrounds the polymeric surfaces. This serves to substantially limit oxygen transport to the cells, retard the diffusion of insulin, and/or deplete the glucose concentration, thereby reducing the utility of the implant. The other variant is intravascular implantation. Islets are cultured on the outside surface of semipermeable tubular membranes housed in a cartridge. The device is implanted in the cardiovascular system as an arteriovenous shunt. Thrombosis is a major problem with this device, but animal experiments using wide-bore tubular mem-branes suggest that long-term patency may eventually be attained without chronic systemic anticoagulation. Another disadvantage is the pre-sence of transport delays in response to glucose stimulation which are associated with diffusion of glucose and insulin through the membrane and into the shell space. One solution to this slow response is to drain the islet chamber, thereby inducing convective transport through ultrafiltration from the bloodstream. However, long-term studies indicate that the ultrafiltration rate declines substantially due to pro-tein deposition on the membrane.

The physiologic environment within hybrid devices may be very different from that to which islets are accustomed in vivo. In the in vivo state, islets are highly vascular-ized by a capillary network which provides for adequate local transport to all cells throughout the islet [2]. In an implanted device this microcirculation no longer func-tions. The islet must depend instead on external diffusion through a series of resistances to supply its metabolic re-quirements and transport the secreted insulin into the vascular system of the host. In the extravascular case, oxygen as well as glucose and other insulin secretagogues must diffuse from the capillary network, through a layer of intervening fibrous tissue which may include cells such as fibroblasts and macrophages, through the membrane, and finally throughout the islet mass. In return, secreted in-sulin as well as other pancreatic hormones, carbon dioxide, and other metabolic wastes must diffuse out through the same series of resistances. As a result, encapsulated islets are potentially subjected to subnormal oxygen tensions, acidic pH, higher than normal levels of insulin and other islet hormones and reduced concentrations of glucose and other insulin secretagogues.

We have studied two problems which deal with the influence of these transport resistances on insulin secretion in hybrid devices. The first is an assessment of the effect of diffusive transport delays on the observed insulin secretory dynamics in an intravascular immunoisolation device [3,4]. The second, described below, concerns the effect of oxygen partial pressure on islet viability and on the intrinsic insulin secretion kinetics of islets.

In any implanted hybrid device, the pO_2 to which islet cells are exposed may be substantially less than that in the nearest blood vessels. To provide a basis for understanding the behavior of an implanted device, we have recently initiated experimental studies to investigate the effect of low pO_2 on (1) the viability of cells in the interior of the islet, and (2) the intrinsic insulin secretion kinetics. These studies are being carried out, respectively with islets (1) in static culture and (2) in a perifusion system.

To provide guidance in experimentation and to aid in interpretation of data, we developed a theoretical model for steady state oxygen diffusion and consumption in an idealized spherical islet. Oxygen consumption kinetics are assumed to follow Michaelis-Menten kinetics. Diffusion and kinetic parameters are estimated from data in the literature. The pO_2 at the islet surface can be specified or an external mass transfer resistance can be interposed. The complete solution is obtained using numerical techniques.

Illustrative results are shown in Figure 1 which is a plot of intra-islet pO_2 profiles for a 240 μm diameter islet exposed to 40 mm Hg at its surface. Under basal conditions (100 mg/dl glucose), the islet is fully oxygenated. When stimulated by 300 mg/dl glucose, the increased O_2 uptake leads to a predicted steeper pO_2 gradient and an anoxic core that may become nonsecreting and/or nonviable over the long run. Similar calculations have shown that a 240 μm islet respiring at basal levels should develop a necrotic core if its surface pO_2 drops below 20 mm Hg.

In order to test predictions of the model, islets of Langerhans were isolated from rats using standard collagenase isolation techniques, then handpicked under a microscope and placed in tissue culture petri dishes containing

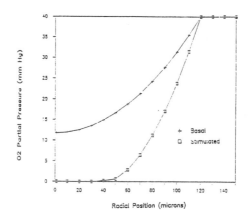

Figure 1.   Profiles of pO_2 estimated from theoretical model for 40 mm Hg
at the surface.   Parameters used in calculations were O_2 diffusion coefficient = 1.5 x $10^{-5}cm^2$/sec, islet diameter = 240μm,
Michaelis constant = 0.2 mm Hg, maximal oxygen uptake rate =
1.9 x 10^{-8} g mol/(cm^3·sec) for basal state, 4.0 x 10^{-8} g mole/
(cm^3·sec) for glucose-stimulated state.

minimum essential Eagles medium with 100 mg/dl glucose, 10%
Newborn Calf Serum, and 100 U/ml Penicillin and 100 mg/ml
Streptomycin.   The petri dishes were placed in temperature
and humidity controlled incubators in which the gas concentration was set at either 33 mm Hg or 133 mm Hg oxygen.
Under these conditions, it was estimated that the surface
pO_2 at the islet periphery was, respectively, 18 or 100 mm
Hg.   The islets were cultured for 40 hours, following which
they were stained with Trypan blue.   In general, the cores
of the islets cultured under low oxygen conditions stained
dark blue, whereas the periphery and the entire volume of
the islets cultured under high oxygen conditions remained
viable (Figure 2).   This data correlates well with the prediction of the theoretical model in that islets exposed to
sufficiently low surface pO_2 levels developed necrotic cores
under basal conditions.   The extent of necrosis is a function of the islet size with larger islets forming a larger
necrotic core, as is also predicted by the model.

In order to determine the effect of low pO_2 on insulin
secretion kinetics, we developed a perifusion system that is
capable of controlling and measuring medium flowrate, pH,
temperature, glucose, and insulin concentrations. pO_2 in
the perifusion medium can be regulated by varying the composition of the gas to which gas permeable silastic tubing,
through which the medium flows, is exposed.   In each run, 10

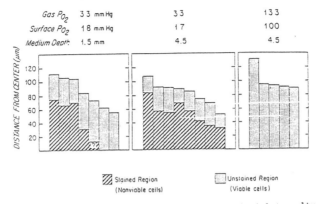

Figure 2. Dimensions of viable and necrotic regions in islets cultured for 40 hr at different values of gas pO₂. Surface pO₂ values were estimated from a theoretical model.

isolated rat islets were handpicked, individually sized, and placed on a 1 μm nitex membrane inside the cell chamber of the perifusion system. The cells were perifused with the same medium as described above containing 100 mg/dl glucose for a period of 40 min under either normal oxygen (140 mm Hg) or reduced (10-40 mm Hg) oxygen conditions. At the end of 40 min, the medium was switched to 300 mg/dl glucose, and fractions were collected at first every minute and later every 3 min for insulin assay. The results of one set of these experiments are shown in Figure 3. The islets perifused with medium equilibrated with 10 mm Hg oxygen released only 59% as much basal insulin as did islets under normal oxygen (140 mm Hg) conditions and only 16% as much under glucose-stimulated conditions. Despite the overall inhibiton, first-phase release was similar under both sets of conditions. Other experiments have indicated that over this time period, normal insulin secretion can be restored if the oxygen is returned to normal levels, indicating that the absence of second-phase secretion is reversible and is not caused by cell death. This is confirmed by the absence of Trypan blue staining in post-perifusion islets.

These initial results suggest that oxygen transport limitations may have profound effects on the viability and insulin secretion capability of living islets contained within implanted hybrid devices. They indicate the need for device designs which ensure an adequate oxygen supply, for example, by close proximity to blood vessels as well as by using materials of construction which minimize fibrotic encapsulation in the case of extravascular implants. In a

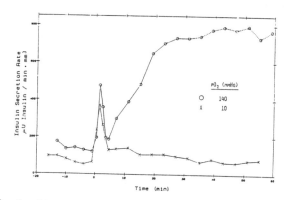

Figure 3. Insulin secretion kinetics in a perifusion system after stimulation by glucose concentration increase from 100 to 300 mg/dl at high (140 mm Hg) and low (10 mm Hg) oxygen conditions. Symbols represent average of two runs. Experimental conditions were medium flowrate = 0.7 ml/min, temperature = 37°C, mean islet diameter = 200μm.

broader sense, these results may have implications for oxygen depletion effects on synthetic or secretory functions in tissues or cell aggregates used in other therapeutic or biotechnological applications.

REFERENCES

1. Scharp, D.W., Mason, N.S., and Sparks, R.E. (1984). Islet Islet immunoisolation: The use of hybrid artificial organs to prevent islet tissue rejection. World J. Surg. 8:221-229.
2. Bonner-Weir, S., and Orci, S. (1982). New perspectives on the microvascular of the islets of Langerhans in the rat. Diabetes 31:883-339.
3. Colton, C.K., Solomon, B.A., Galletti, P.M., Richardson, P.D., Takahashi, C., Naber, S.P., and Chick, W.L. (1980). Development of novel semipermeable tubular membranes for a hybrid artificial pancreas. In Ultrafiltration Membranes and Applications, A.R. Cooper, ed., Plenum Press, N.Y., pp. 541-555.
4. Colton, C.K., and Weinless, N.L. (1987). A theoretical model for insulin secretion rate in a hybrid artificial pancreas. In Artificial Organs, J.D. Andrade et al., eds., VCH Publishers, N.Y., pp. 641-655.

Tissue Engineering, pages 223–230
© 1988 Alan R. Liss, Inc.

COMPARISON OF ARTIFICIAL PANCREAS DEVICES

Robert A. Peura

Biomedical Engineering Program
Worcester Polytechnic Institute
Worcester, Massachusetts 01609

Diabetes is a major health problem affecting a large segment of our population. Conventional administration of insulin using subcutaneous injections is not comparable to the complex hormonal feedback regulatory system. Many attempts have been made to develop artificial pancreas systems which mimic the homeostatic system of the body. This paper reviews these approaches and discusses their advantages and disadvantages.

INTRODUCTION

Glucose is the main circulating carbohydrate in the body. In normal fasting individuals, the concentration of glucose in blood is very tightly regulated, usually in the range between 80 and 90 mg./100 ml., during the first hour or so following a meal. The hormone insulin, normally produced by the pancreas' beta cells, promotes glucose transport into skeletal muscle and adipose tissue. In diabetes mellitus, insulin regulated glucose uptake by the cells is compromised and, consequently, blood glucose can elevate to abnormal concentrations ranging from 300 to 700 mg./100 ml.

Chronic long-term complications associated with diabetes include the development of coronary artery disease, visual impairment, renal failure, peripheral vascular disease, etc. It has been estimated that in the U.S. the total economic impact of diabetes is in the order of 14 billion dollars.

The treatment of diabetes with prescribed injections of insulin subcutaneously, results in inadequate control of glycemia compared to normal homeostatic control. Blood glucose levels rise and fall several times a day. Normoglycemia, using an "open-loop" insulin delivery approach, is difficult to maintain. It is apparent that improved diabetes management is needed. Several therapeutic alternative approaches have been proposed to optimize blood glucose control. These approaches range from: total or segmental pancreas transplantation; islet transplantation; encapsulated islets; implantation of polymeric capsules; and insulin pump delivery systems.

PANCREATIC TRANSPLANTATION

More than 1000 pancreatic transplants (whole or segmental) have been performed worldwide. Eight years is the longest period for a pancreatic transplant patient to remain insulin-independent. Segmental grafts are obtained from living related donors or cadavers. Pancreatic transplants are presently restricted to diabetic patients whose secondary diabetic complications are, or those anticipated, more serious than the likely life time anti-rejection therapy side effects. The availability of donor pancreases as a source of islets for grafting appears to be a challenge. However, there are 5,000 kidney transplants per year in the U.S. and approximately 10,000 new type I diabetic persons are diagnosed per year, with 50% of them developing serious complications. Thus, the donor procurement for pancreas transplantation should not be any more difficult than for kidney transplantation. [1]

A major technical problem in either the segmental or whole pancreatic transplant is managing the exocrine secretion produced by the transplanted tissue. A large portion of the failure rate in either the combined pancreas and kidney or single pancreas transplant procedures has been mortality due to patient selection. Graft thrombosis has been indicated as a primary cause of graft loss. The use of anticoagulation therapy has reduced the incidence of this problem. The most significant problem of pancreatic transplantation is chronic rejection occurring within two years. The survival rates for major centers in the U.S. and Europe are approaching 60% graft success. Patient survival, with careful patient selection, is 90% at two years. It appears that successful transplant therapy can prevent the development of secondary complications of diabetes if not, reverse them. [1]

PRE-ISLET CELL TRANSPLANTATION

Proliferated human cells have been transplanted into immuno-suppressed diabetic patients who are undergoing or have recently received a kidney transplant. These cells are usually inserted under the transplanted kidney capsule. Fetal progenitor cells are identified and isolated using monoclonal antibodies. Unwanted cell material, including fibroblasts, which cause transplant rejection are eliminated by a several stage purification process. The cells are grown and proliferated using a serum-free medium in a controlled growth environment. Once the pre-islet cells are placed in the body they vascularize, mature into islet cells and begin secreting insulin as a function of the glucose levels present. [2,3]

Studies in animals have demonstrated functional equivalence to non-proliferated transplanted cells. Human clinical studies are currently underway in which cells are transplanted into patients who are undergoing or who have recently received a kidney transplant and are thus undergoing immunosuppression therapy.

There is a potential for mass production of these cells. Large quantities of these cells may potentially be available for transplantation. Presently, immunosuppression therapy is necessary, however, it is proposed that highly purified cells will reduce the possibility of transplant rejection.

ENCAPSULATED ISLETS

Encapsulated islets perform normal graft functions without the use of immunosuppression since the grafted cells are protected with an artificial membrane which prevents an immune rejection. The capsule membrane is semipermeable such that it is permeable to insulin and glucose, but excludes antibodies which would otherwise cause cell rejection. [4]

These islet encapsulated systems can be either macroencapsulated or microencapsulated. These terms differentiate the physical arrangement by which the donor islet cells are isolated from the hose immune system. Current macroencapsulation techniques grow islet cells on hollow fibers which serve as artificial capillaries. These fibers, similar to artificial kidney fibers, are perfused with blood by means of a vascular shunt. The fibers are permeable to glucose and insulin, but essentially impermeable to larger molecules such as gamma globulin. Localized anticoagulation systems for the intravascular device must be developed in order to allow for long term patency without the side effects of systemic anticoagulation.

Other investigators have used intraperitoneal implantation of macroencapsulated islet cells. Altman et al [5] prepared 200 human islet cells from a benign insulinoma and placed them in Amicon XM50, permselective tubular membranes. This was placed intraperitoneally as a free floating implant. It was found that glucose insulin metabolism was normalized in 1/2 the diabetic rats studied. Limitations with this device include: the fragility of the semipermeable membrane (Amicon fiber); limited diffusion surface area; fibrotic reaction;and the destruction of normal cells within the fiber after long term implantation.

The two major problems of vascular complications and diffusion dynamics are minimized by microencapsulating individual islet cells with a biocompatible, semipermeable membrane. This allows diffusion of glucose and insulin while isolating the islets from the host immune system. Rat islets have been encapsulated in biocompatible alginate-polylysine-alginate membranes. Micro-encapsulated allotransplants of rat islets reversed diabetes in diabetic rats. The longest allotransplant survived 780 days with an intact capsule of functioning islet cells. Additional studies dealing with large animals and the development of techniques to secure islets on a mass scale, must be completed before clinical trials might begin. [6].

CONTROLLED RELEASE POLYMERIC INSULIN SYSTEMS

Langer has summarized four types of polymeric insulin delivery systems. [7]. These devices, in various stages of development, include: 1) constant rate of insulin delivery from nondegradable polymers; 2) insulin delivery from degradable polymers; 3) variable insulin delivery from polymeric systems controlled by magnetic or ultrasonic forces; and 4) variable insulin delivery dependent upon a reaction caused by glucose within the polymer. Langer's group developed a method which provided constant insulin release using a non-degradable insulin-polymer pellet system. This produced normoglycemia in diabetic rats for a period of 100 days. Insulin encapulated biodegradable polymers have been tested in diabetic rats with resultant release times from a few days up to a month.

Polymeric systems have been designed to increase the release of insulin on demand for meal time bolus applications. Increased insulin release has been achieved with the application of: 1) a magnetic field to a nondegradable elastic polymeric matrix in which small magnetic beads are incorporated with insulin; and 2) the application of ultrasound on nondegradable polymers. [7].

Several methods for self-regulated polymeric membrane insulin systems have been explored. Horbett, et al developed a glucose sensitive membrane which senses the level of glucose and controls the delivery of insulin. [8]. The polymer membrane for glucose delivery works as follows. Glucose diffuses into a synthetic polymer network and reacts with glucose oxidase enzyme. The conversion of glucose into gluconic acid in the membrane protonates basic functional groups in the membrane. Consequently, electro-static repulsion occurs which results in increased membrane swelling. This increase in permeability is used to deliver insulin in response to the presence of glucose.

Polymeric capsules are in a pre-clinical state. They have a potential for use as self-regulated insulin delivery systems. Most membrane materials, suitable for constructing a polymer network, have been shown to be biocompatible. Questions still exist concerning the long-term stability of the polymeric system, enzyme entrapment, response time, insulin membrane permeability and the effect of physiological buffering. Another advantage is that insulin is stored in powdered form and thus is less likely to aggregate than when it is in solution. The small size, ease of implantation, and low cost are also advantages of the polymeric systems. [9].

INSULIN PUMPS

External and implantable insulin pumps are currently being used for insulin delivery in the treatment of diabetes. External portable pumps are worn on the body using various harnesses or belts and deliver insulin subcutaneously through small needles or cannulae. Implantable pumps make use of either intraperitoneal or intravenous delivery routes. One problem with insulin pumps has been the tendency of commercially available insulin to clog the insulin delivery system. A new glycerol/insulin formulation, which should minimize the insulin precipitation problem, is in clinical trials.

The advantages of insulin pump therapy include: 1) devices are being used by selected diabetic patients; 2) commercial manufacturers produce an adequate supply; and 3) immunosuppression therapy is not needed.

Glucose Sensors

The lack of a viable long term glucose sensor limits the ability of insulin pumps to self-regulate their output. A solution to this problem would be to "close the loop" using a self-adapting insulin infusion device with a glucose controlled biosensor, which could continuously sense the need for dispensing insulin at the correct rate and time. Unfortunately, present day glucose sensors do not meet these stringent requirements. Although significant progress has been made toward the development of various novel glucose sensors, further advances must be made before these sensors can be perfected and adapted for in vivo monitoring of glucose. (10)

Shichiri has developed a miniature needle glucose oxidase electrode for measuring glucose levels in subcutaneous tissues. The functional life of the electrode is seven days. Normoglycemic control has been achieved in dog studies. [11] Shultz has developed an optical affinity sensor which is being tested in laboratory animals by a commercial firm. [12] The work of Glaser et al showed that it is feasible to measure blood glucose noninvasively by monitoring the glucose levels on the rabbit palpebral conjunctiva surface. [13]

Substantial progress has been made on the development of insulin pump systems. However, the basic limitation in an insulin self-regulating pump is the biosensor. Significant work still remains to be done in this area.

<u>SUMMARY</u>

This paper has discussed a number of innovative approaches to the control of blood glucose in diabetic individuals. Many challenging problems exist along each pathway. The needs are great and the rewards will be many for a viable solution to this problem.

REFERENCES

1. Sutherland, D.E.R. (1986). Current status of pancreas transplantation. Clinical Diabetes 4:55-70.

2. Gray, D.W.R. and Morris, P.J.
    (1987).Developments in isolated pancreatic
    islet transplantation.   Transplantation
    43:321-331.
3. Hu Yuan-feng, Zhang Hong, Zhang Hong-de,
    Shao, An-hua, Li Li-xian, Zhou Hui-qing,
    Zhao Bao-Hua, Zhou Yu-gu (1985) Culture of
    human fetal pancreas and islet trans-
    plantation in 24 patients with type 1
    diabetes mellitus. Chinese Medical
    Journal 98(4):236-243.
4. Araki, Y., Solomon, B. A., Basile, R. M. and
    Chick, W.L. (1985). Biohybrid artificial
    pancreas; long-term insulin secretion
    by devices seeded with canine islets.
    Diabetes, 34:850-854.
5. Altman, J.J., Houlbert, D., Chollier, A.,
    Leduc, A., McMillan, P., and Galletti,
    P.M. (1984). Encapsulated human islet trans-
    plants in diabetic rats. Trans.Am.Soc.
    Artif.Intern.Organs Vol.XXX:382-386.
6. Sun, A. M. (1987). Encapsulated versus
    modified endocrine cells for organ
    replacement. Trans.Am.Soc.Artif.Intern.Organs
    Vol.XXXIII:787-790.
7. Buchwald, H. (1987). Insulin replacement:
    bionic or natural. Trans.Am.Soc.Artif.
    Intern. Organs VolXXXIII:806-812.
8. Horbett, T.A., et al (1984). A bioresponsive
    membrane for insulin delivery. Proc.
    Univ. Utah Int. Symp. Recent Adv. in
    Drug Del. Sys.
10. Segawa, M., Nakano, H., Nakajima, Y., Murao,
    Y., Nakagawa, K., and Shiratori, T. (1987).
    Effect of hybrid artificial pancreas on
    glucose regulation in diabetic dogs.
    Transplantation Proceedings, XIX:985-988.
11. Peura, R.A. and Mendelson, Y. (1984).
    Blood glucose sensors: an overview.
    IEEE/NSF Symposium on Biosensors, Los
    Angeles, CA Conference Proceedings 63-69.
12. Guilbeau, E.J., Clark, L.C., Pizziconi, V.
    B., Schultz, J.S., and Towe, B.C.
    (1987). Trans Am Soc Artif. Intern.
    Organs, XXXIII:834-837.
13. Glaser, G., Peura, R.A., Mendelson, Y., and
    Shahnarian, A. (1987). Feasibility of the
    palpebral conjunctiva as a site for
    continuous, noninvasive glucose monitoring.
    Proceedings IEEE Engineering in Medicine &
    Biology Society, 9:786-787.

Tissue Engineering, pages 231–239
© 1988 Alan R. Liss, Inc.

CONTROL OF DISEASE RECURRENCE FOLLOWING ISLET GRAFTING INTO SPONTANEOUSLY DIABETIC NOD MICE*

Yi Wang and Kevin J. Lafferty

Barbara Davis Center for Childhood Diabetes
Department of Microbiology/Immunology & Pediatrics
University of Colorado Health Science Center
Denver, CO 80262

INTRODUCTION

Type I diabetes in man is an immunologically mediated disease. The disease also has an immunological etiology in a rodent models, the NOD mouse(1,2,3). In this model the disease is T cell dependent (3), and evidence from our laboratory suggests that the disease is mediated by a CD4 effector T cell(4). These findings have lead us to suggest that islet damage is the result of inflammatory tissue damage, and that the specificity of the disease process – the specific destruction of ß cells within the islets – results from the particular sensitivity of islet ß cells to free radical damage (5,6).

MATERIALS AND METHODS

Animals.

NOD/Den mice were derived from a nucleus obtained from Ehime University, Japan, and have been inbred in Denver for seven generations. The incidence of diabetes in our colony has been estimated 77% in females and 20% in males. NOD mice were screened for diabetes and maintained by

*Grant support: This work was supported in part by NIH grant DK 33470

insulin therapy as previously described (4). Male
6-8 week old BALB/c mice used as the donors of
islet and pituitary tissue were purchased from the
Jackson Laboratory(Bar Harbor,ME).

Experimental Protocol and Groups.

        NOD mice were divided into 5 groups in this
study. Group 1 was made of diabetic NOD mice. All
animals in this group were transplanted with
cultured BALB/c islet and pituitary tissue on day
0. Group 2 were made of non-diabetic NOD
recipients and were transplanted with cultured
islet tissue on day 0. Animals in group 3 were
non-diabetic NOD recipients and were transplanted
with cultured BALB/c pituitary tissue on day 0.
Animals were injected intraperitoneally with 2 x
10^6 donor spleen cells at the time of grafting.
Animals in group 4 and 5 were diabetic NOD
recipients. Group 4 were subjected to CD4 T-cell
depletion by injecting intraperitoneally (IP) with
monoclonal antibody GK1.5 (anti-CD4) (7): 200mg/kg
on days -14, -7, 0 and were transplanted with
BALB/c islet on day 0. Group 5 were subjected to
CD8 T-cell depletion by in vivo administration
(IP) of monoclonal antibody 116-13.1 (anti-CD8)
(8) 200mg/kg on days -3, +4, relative to the time
of grafting islet tissue on day 0.

Preparation, Culture and Grafting of Islet and
Pituitary Tissue.

        Islet tissue was prepared from the pancreas
of BALB/c mice and was cultured for 7 days in an
atmosphere of 95% O2, 5% CO2 as described in
detail previously (9). Nine clusters of cultured
BALB/c islet, each containing 50-60 islets were
transplanted under the kidney capsule of test
animals on day 0. The pituitary tissue was
cultured in RPMI 1640 medium supplemented with 10%
fetal calf serum, in 95% O_2 and 5% CO_2 for two
weeks and was transplanted in the same manner as
the islets tissue (9). Following transplantation,

blood glucose was measured once a week. The animals were sacrificed and the grafts were removed for histological examination when the experiments were completed.

Preparation GK1.5 and 116-13.1 Antibody.

Ascitic fluid containing monoclonal antibody specific for CD4 T cells (GK1.5), or specific for CD8 T cells (116-13.1) was obtained from sublethally irradiated BALB/c mice as previously described (4).

Analysis of CD4 & CD8 T Cells in Peripheral Blood.

The level of CD4 and CD8 T cells as a percentage of total peripheral lymphocytes and cellular kinetics after anti-CD4 or anti-CD8 treatment were studied by flow cytometric analysis as described previously (4).

RESULTS

Disease Recurrence in Islet Tissue Transplanted to Diabetic NOD Mice Provides a Model for Study of the Pathogenesis of the Disease Process.

Culture of islet and pituitary tissue in an oxygen rich atmosphere prior to grafting eliminates active antigen presenting cells (APCs) from the graft, and allows allotransplantation without immunosuppression (10). Grafting of cultured islet tissue to diabetic NOD mice provides a model for the study of the disease process. Diabetic NOD mice transplanted with cultured BALB/c islets and pituitary tissue destroy the islets but not the pituitary graft (Table 1). On the other hand, when the recipient NOD mice were challenged with spleen cells of donor origin the pituitary graft was promptly destroyed (Table 1). Histological examination of

the grafted tissue in diabetic NOD recipients at
various times post-transplantation showed well
granulated islets 4 days after grafting. However,
by 7 days an intense mononuclear-cell infiltration
was seen in and around the islet tissue which by
this stage was degranulated (unpublished data).
Fourteen days post-transplantation, all animals
remained hyperglycemic, and histological
examination revealed that the islet tissue was
totally destroyed, little or no damage was seen in
cultured pituitaries from the same donors
transplanted along with the islet tissue (Table
1). When cultured BALB/c islet tissue was
transplanted to non-diabetic NOD mice, little or
no infiltration was seen around the islet tissue,
and the islets were well granulated when examined
histologically 28 days post-transplantation (Table
1).The tissue specificity of the pathological
process observed when cultured BALB/c islet tissue
was grafted to diabetic NOD mice and its
association with diabetes in the recipient animal,
provides strong evidence that this destructive
process represents disease recurrence in the
grafted tissue, and is not the result of an
allograft response.

TABLE 1
ALLOGENEIC ISLET GRAFTS IN DIABETIC NOD MICE ARE
DESTROYED BY DISEASE RECURRENCE

Group#	Islet Donor	Recipient	# of grafts surviving [a] Islet	Pituitary
1	B/C	D-NOD	0/6	6/6
2	B/C	ND-NOD [b]	4/4	N.T.[c]
3	B/C	ND-NOD-primed [d]	N.T.	0/3

[a]diabetic (D-) or Non-diabetic (ND-) NOD
recipients were transplanted with cultured BALB/c
(B/C) islet and pituitary on day 0. Graft survival
was monitored by macroscopic and histological
examination 28 days post transplantation and, in
addition, by blood glucose measurement in group 1.

Depletion of CD4 or CD8 T Cells by In Vivo
Administration of Specific Antibodies.

In vivo administration of GK1.5, 200mg/kg on
days -14, -7 and day 0 reduced CD4 T cells in the
periphery from the initial levels of approximately
39% of total lymphocytes to less than 3 % by day
0. CD4 T cells remained at this low level for
another week after the cessation of the GK1.5
treatment. With the further passage of time there
was a significant rise of the CD4 T cell level,
which reached approximately half the level seen in
the animals prior GK1.5 treatment, by 50 days
after treatment (4). The decrease in CD4 T cells
in peripheral blood was accompanied by an increase
in percentage of CD8 T cells on day 0 (From 23.6%
to 41.1%). Similarly, the in vivo administration
of monoclonal antibody 116-13.1(anti-CD8),
200mg/kg on days -3, +4, reduced CD8 T cells in
the peripheral blood from the initial levels of
approximately 24% to less than 2 % by day 0, and
remained at this low level for at least two weeks.
In this case, the percentage of CD4 T cells
increased from 38.6% to 50.9% after the reduction
of CD8 T cells in peripheral blood on day 0. In
summary, in vivo administration of anti-CD4
antibody (GK1.5) or anti-CD8 antibody (116-13.1)
specifically deplete CD4 or CD8 T cells. This
approach allowed us to test which T cell subset
were required for the expression of the disease
process in diabetic NOD mice.

Requirement of CD4 and CD8 T Cells for the
Expression of Disease.

Transplantation of cultured BALB/c islet
tissue to spontaneously diabetic NOD mice failed
to bring the blood glucose into normal range. This

[b]Male non-diabetic NOD mice which were 4-5
month old were monitored once a week for the
absence of glycosuria and hyperglycemia.
[c]N.T.: not tested.
[d]Animals in group 3 were injected (IP) with 2
x 10[6] BALB/c spleen cells at the time of grafting.

inability of islet tissue to function results from
disease recurrence in the grafted tissue (Table
1). When the cultured islet tissue was
transplanted to animals treated with the anti-CD4
antibody (GK1.5) 200mg/kg on days -14, -7 and 0
(Group 4), all grafts brought the blood sugar to
the normal range within a week of transplantation
(Table 2). However, normoglycemia was not
maintained indefinitely; grafted animals became
hyperglycemic 10-30 days post transplantation.
The return of the disease process was correlated
with the reappearance of CD4 T cells in the
peripheral blood. The average percentage of CD4 T
cell at the time of disease recurrence was 18.2
+/- 2.7% (n=5). On the other hand, when the
cultured islet tissue was transplanted to the
animals treated with the anti-CD8 monoclonal
antibody (116-13.1) 200mg/kg on day -3, +4 (Group
5), grafts failed to bring the blood glucose into
normal range and all the animals remain
hyperglycemic (Table 2).

TABLE 2
DISEASE RECURRENCE IN BALB/c ISLETS GRAFTED TO
DIABETIC NOD MICE IS CD4 T CELLS DEPENDENT

Group #	Antibody Pre-treatment	Graft Survival Time [a]
1	None	<14,<14,<14,<14,<14,<14
4	Anti-CD4 [b]	<14, 30, 29, 21, 28, 28
5	Anti-CD8 [c]	<14,<14,<14,<14,<14,<14

[a]Diabetic NOD mice were transplanted with
cultured BALB/c islet tissue on day 0. Graft
survival time is scored as the days of
consecutive blood sugar readings in normal range
(6.3 ± 3.1 mmol / l).
    [b]Animals in group 4 were treated with GK1.5
(anti-CD4) 200mg/kg on days -14, -7, 0.
    [c]Animals in group 5 were treated with
116-13.1 (anti-CD8) 200mg/kg on days -3 and +4.

DISCUSSION

The first set of experiments demonstrates that disease recurrence in cultured allogeneic islet tissue transplanted to diabetic NOD mice provides a model system for the study of the pathogenesis of type 1 diabetes. The immunogenicity of islet tissue was reduced by culturing in an oxygen-rich atmosphere, which allows transplantation of cultured tissue across major MHC barrier without immunosuppressive therapy (10). When cultured islet and pituitary tissue were simultaneously transplanted to diabetic NOD mice, the pituitary tissue remained intact but islet tissue was totally destroyed by 14 days post transplantation. The pituitary is a sensitive indicator of allograft immunity and was destroyed following active immunization of the recipients (Table 1). On the other hand, the same islet allografts remained intact in nondiabetic NOD recipients (Table 1). The tissue specific islet destruction associated with disease process in the recipients lead us to conclude that islet tissue was destroyed by disease recurrence in the graft.

Our second set of experiments demonstrates that monoclonal antibodies GK1.5 (anti-CD4) and 116-13.1(anti-CD8) specifically deplete CD4 and CD8 T cells in the peripheral blood. By in vivo administration of antibody GK1.5 or 116-13.1 to deplete CD4 or CD8 T cells prior to the transplantation of islet tissue, we determined which T cell subsets were required for the expression of disease recurrence. Our results show that disease recurrence in the graft was delayed significantly by depletion of CD4 T cells, indicating that CD4 T cells were required for the expression of disease process; depletion of CD8 T cells did not affect expression of the disease process. In a separate study, we have evidence that this same anti-CD8 antibody is biological active (unpublished data). These results contrast with those of Wicker ' group who found both CD4 and CD8 T cells are required for initiation of disease in young non-diabetic NOD mice (11). We have confirmed these findings in a disease

transfer model (unpublished data). We propose that
some form of cooperative action between CD4 and
CD8 T cells required to initiate disease process,
but that only CD4 T cell is required for the
expression of disease. Our findings suggest that
the CD4 T cell is the immunological effector in
the process of islet destruction and that islet
damage is the result of inflammatory tissue
damage. Specific ß cell destruction could result
from a nonspecific inflammatory reaction because
the particular sensitivity of islet ß cells to
free radical damage (5,6). Another possibility is
that IL1 produced in the inflammatory is the cause
of beta cell damage (12). These findings take us a
step forward in understanding the pathogenesis of
type I diabetes in this animals model and open up
a new approach to the control of the disease
process.

SUMMARY

     Methods for isolation and transplantation of
allogeneic islet tissue without immunosuppression
has been developed. When islet tissue was
transplanted into spontaneous diabetic NOD mice,
the tissue were acutely destroyed by disease
recurrence in the graft. The disease recurrence is
a CD4 T lymphocytes dependent process. These
findings, in addition to the finding that radical
scavengers can inhibit disease development in the
grafted tissue, suggest that disease results from
an inflammatory process, and that free radicals
generated at the site of inflammation are directly
involved in the islet ß cell destruction.

REFERENCES

1. Miyazaki A, Hanafusa T, Yamada K, Miyagawa J,
Fujino-Kurihara H, Nakajima H, Nonaka K, and Tarui
S (1985). Predominant of T lymphocytes in
pancreatic islet and spleen of pre-diabetic
non-obese diabetic (NOD) mice: a longitudinal
study. Clin Exp Immunol 60: 622.
2. Mori Y, Suko M, Okudaira H, Matsuba I, Tsuruoka
A, Sasaki A, Yokoyama H, Tanase T, Shida T,

Nishimura M, Terada E and Ikeda Y (1986).
Preventive effects of cyclosporine on diabetes in
NOD mice. Diabetologia 29: 244.
3. Wicker LS, Miller BJ, Mullen Y (1986). Transfer
of Autoimmune Diabetes Mellitus With Splenocyte
From Nonobese Diabetes (NOD) Mice. Diabetes 35:
855.
4. Wang Y, Hao L, Gill R, Lafferty KJ (1987).
Autoimmune Diabetes in the NOD mouse is L3T4
T-lymphocytes Dependent. Diabetes 36: 535.
5. Nomikos IN, Prowse SJ, Carotenuto P, Lafferty
KJ (1986). Combined Treatment With Nicotinamide
and Desferrioxzmine Prevents Islet Allograft
Destruction in NOD mice. Diabetes 36: 1302.
6. Malaisse WJ (1982). Commentary Alloxan Toxicity
To Pancreatic B-Cell. A New Hypothesis. Biochem
Pharmac 31: 3527-34.
7. Dialynas DP, Wilde DB, Marrack P, Pierres A,
Wall KA, Havran W, Otten G, Loken MR, Pierres M,
Kappler J, and Fitch FW (1983). Characterization
of the Murine Antigenic Determinant, Designated
L3T4a, Recognized by Monoclonal Antibody GK1.5:
Expression of L3T4a by Functional T Cell Clones
Appears to Correlate primarily with Class II MHC
Antigen-Reactivity. Immunological Rev 74: 29.
8. Shen FW et al (1981). In Hammerling GJ et al
(eds) "Monoclonal Antibodies and T-Cell
Hybridomas". Elsevier / North Holland, p 25.
9. Prowse SJ, Simeonovic CJ, Lafferty KJ, Bond BC,
Magi CE, Mackie D (1984). Allogenic Islet
Transplantation Without Recipient
Immunosuppression. In Larner J, Pohl S (eds):
"Methods in Diabetes Research, vol I: Laboratory
Methods, Part A," John Wiley & Sons, p 253.
10. Lafferty KJ, Prowse SJ, Simeonovic CJ (1983).
Immunobiology of Tissue Transplantation: A Return
to the Passenger Leukocyte Concept. Ann Rev
Immunol 1: 143.
11. Miller BJ, Appel MC, O'Neil JJ, and Wicker LS
(1988). Both the Lyt2[+] and L3T4[+] T cell Subsets
Are Required for the Transfer of Diabetes in
Nonobese Diabetic Mice. J Immunology 140: 52.
12. Mandrup-Poulsen T, Bendtzen K, Nerup J,
Dinarello CA, Svenson M, Nielsen JH (1986).
Affinity-purified human interleukin 1 is cytotoxic
to isolated islets of Langerhans. Diabetologia 29:
63.

Tissue Engineering, pages 241–242
© 1988 Alan R. Liss, Inc.

Summary/Discussion

Transplants and Artificial Organs

Bruce E. Jarrell

Dr. Robert Peura gave an overview of artificial pancreatic endocrine replacement. The process of islet cell culture with depletion of dendritic and
other Ia-bearing cells is a promising area. An important drawback is the
need for large numbers of cells; a major advance might occur if islet cells
could be induced to undergo replication. It would appear that islet culture
with depletion of dendritic cells does improve the longevity of islet cells
implanted into experimental animals, suggesting that islet cells themselves
are not potent antigenic stimulants. The process of microencapsulation as
described by Dr. Anthony Lim was presented. There are numerous unresolved engineering issues with respect to the transport of materials across
these membranes and the viability of islets within the membranes.

The use of artificial glucose sensors was also discussed. A major issue
remains the ability to adequately sense the blood glucose level in a timely
fashion and produce an appropriate insulin response to that glucose level.
Surfaces with immobilized glucose oxidase or other enzymes have not
produced long-term acceptable results. Several new types of sensing
devices were discussed, but further investigation is needed in this area.

Dr. Pierre Galletti discussed experiences with the artificial pancreas. Islets
were attached to the surface of a hollow fiber and exposed to glucose-
containing solutions upon the other surface of the fiber. Insulin secretion
occurred in response to a glucose challenge, although in a slightly delayed
fashion. Islet cell attachment to this type of surface is currently poorly
understood. Mass transport of solutes across these hollow fiber membranes
also needs further in depth investigation.

Dr. Clark Coulton discussed pancreatic cellular metabolism. In his experimental setting, islets can successfully be maintained *ex vivo* and respond to

glucose challenges. It was noted that cells within the central part of the islet sphere of cells tend to become ischemic, with blunted insulin response to glucose challenges. If the ambient oxygen is increased to higher levels, the oxygen tension within the center of this mass of cells increases. This is associated with an improved insulin response to a glucose challenge. Oxygen transport may be one limiting factor affecting islet function.

Summary

With respect to pancreatic replacement therapy, currently the only success-ful islet cell replacement in humans has been with partial or total body pancreatic replacement as a transplantation procedure requiring immuno-suppression. The use of individual cells for this replacement is hampered by inability to culture these cells, lack of access to suitable numbers of human cells, and incomplete knowledge regarding the human immunologi-cal response to islet cells of different species implanted into the human. A major issue with living cells is the transport of solutes as well as oxygen such that the cell is able to maintain normal metabolic function. It is possible to attach these cells to a surface and thus allow them to function in close proximity to the bloodstream. Further work in this area, particularly with cellular co-culture, may be useful.

Tissue Engineering, pages 243–247
© 1988 Alan R. Liss, Inc.

VI. Restoration and Maintenance of Neurological Function

Introduction. . . . . William Freed

Even in prehistoric times, mankind attempted to manipulate the functioning of the brain. Trephining, the production of holes in the skull, was employed presumably to allow for the departure of evil spirits. In modern times, brain manipulation has been accomplished primarily through lesioning or stimulation of specific brain regions, such as thalamotomy for Parkinson's disease, frontal lobotomy for schizophrenia, temporal lobectomy for epilepsy, and various experimental forms of localized brain stimulation. In the best cases, such as temporal lobectomy for epilepsy, there has been an anatomical relationship between manifestations of the underlying disorder and the therapy. In other examples, however, the relationship between disorder and therapy has not been clear. During the past 20 years, brain manipulation has shifted toward well-defined neuronal deficits where the therapy is clearly directed at the restoration of normal neuronal functioning. In large part, this has become possible because neurochemically defined functional circuits have been mapped out for several parts of the nervous system. In a few cases, disorders have been specifically linked to damaged neuronal circuits. This has made possible a more precise kind of neuronal manipulation involving attempts to directly influence those systems that are known to be deranged. Nevertheless, as this work progresses, it becomes clear that the manipulations of man upon the brain are at best only able to approximate natural biological processes. To expect to restore complex neuronal systems to their original condition through artificial manipulation is, in most cases, unrealistic.

One central system, the nigrostriatal dopamine system, is known to be damaged in patients with Parkinson's disease, and this malfunction is thought to be responsible for many of the manifestations of the disorder. A number of studies have been directed at finding methods of repairing this system. In these studies, the primary goal has been to find techniques for delivering dopamine in a biologically useful manner directly to the striatum. Methods for accomplishing this have included intercerebral dopamine infusion pumps, transplantation of embryonic dopamine-containing cells into or near the stratum, and transplantation of alternative cell types, such as adrenal medulla or dopamine-producing tumor cells. Each of these methods has potential advantages as well as drawbacks. For example, the question of whether it would ever be possible to transplant tumor cells into a human patient, despite various safeguards, was raised. Delivery of dopamine to the brain from infusion pumps depends entirely upon diffusion, a relatively ineffective means of drug distribution. Embryonic substantia nigra neurons could be subject to damage through whatever process was originally responsible for the disease. Adrenal chromaffin cells suffer only minimal damage in Parkinson;s disease, but these cells do not form direct connections with host neurons following transplantation, working instead through less well understood processes. Even the connections between transplanted embryonic neurons and host brain are incomplete and differ substantially from the connections between normal substantia nigra and striatum. Adrenal medulla autografts have already been performed in a number of patients with Parkinson's disease. Early results are promising, and suggest that these grafts produce some improvement in at least a subset of drug-resistant patients. Nevertheless, there is as yet no conclusive evidence that these grafts have survived, and the mechanism of action of adrenal medulla grafts in human subjects is not completely understood.

Another well-defined form of neuronal damage is spinal cord injury. Here the failure of recovery is clearly related to the absence of regeneration of spinal cord neurites across the injury site. For many years it has been recognized that regeneration of spinal cord neurites can be enhanced by a variety of manipulations that decrease the density of scar formation. Nevertheless, regeneration across the gap was not achieved. Recently, Albert Aguayo and his colleagues have shown that it is possible to promote the growth of spinal cord neurites for long distances by means of implantation of segments of peripheral nerve as "bridges." These studies have clearly shown that the peripheral nerve supporting elements are capable of greatly enhancing the growth of central axons. Nevertheless, this increased neurite growth is not always accompanied by appropriate connectivity of these axons once they exit the distal end of the bridge, and continued growth from the distal end of the bridge is not usually enhanced. Kliot has described a device based on an astrocyte-coated millipore filter which very effectively promotes regeneration of dorsal root sensory axons into spinal cord grey matter. Very preliminary data suggested that these implants produced some functional effects. Whether these regenerating axons produced functionally effective connections with target neurons remains, however, to be determined.

In contrast to spinal cord, peripheral nerve axons regenerate readily, even when the nerve is completely severed. In this case, however, adequate functional recovery often does not occur because of misrouting of axons at the point of reconnection. Even very precise microsurgical repair often produces an unsatisfactory functional outcome. When peripheral nerve is crushed, rather than severed, functional recovery almost always occurs. Even though the axons are destroyed, enough of the surrounding structure (primarily basal lamina) remains to guide regenerating neurites to their correct targets. There has been an interest in the possibility of using synthetic guidance channels to promote the growth of peripheral nerve fibers. Aebischer has studied the properties of polymers (including permeability, surface properties, and electrical charge) used to make guidance channels to determine what most effectively promotes peripheral nerve regeneration across a gap. A guidance channel was developed which promoted the regeneration of peripheral nerve across a very wide gap. This provides valuable information about the properties of peripheral nerve sheath which most effectively promote regrowth. Insofar as clinical application is concerned, one possibility is that guidance channels could be employed to traverse large areas of tissue damage where the original nerve is lost. Since this is a relatively unusual circumstance, a more important unresolved issue is the degree to which regenerating neurites are able to employ intrinsic trophic guidance mechanisms during growth through these guidance channels in order to develop the appropriate functional connections. It has also not been shown that functional recovery occurs more frequently when synthetic guidance channels are employed, as compared to repair by conventional microsuturing methods.

A very promising area of investigation is the control of neuronal development and growth by chemical substances. There is a possibility that such substances could be employed to promote functional recovery following brain injury. The prototypical neuronal growth factor is Nerve Growth Factor (NGF). NGF is found in high concentrations in male mouse submaxillary gland, and selectively influences peripheral sympathetic neurons, peripheral embryonic sensory neurons, and central cholinergic neurons.

NGF can promote both neuronal survival and neurite extension, whereas most other similar substances have only one or the other type of activity. There are few known cases of substances which inhibit neurite extension. Cunningham and his colleagues report that thrombin causes the retraction of neurites from neuronal cell cultures. Protease Nexin I, a potent thrombin inhibitor, stimulates neurite extension through inhibition of thrombin activity. Thrombin may, therefore, act as an endogenous inhibitor of neuronal process extension. These data further suggest that neuronal differentiation is subject to control through a complex system of inhibitory and stimulatory factors. Further understanding of the circumstances under which these factors are active may contribute to the development of methods for influencing the outcome of neuronal damage.

The secretory properties of neuronal and endocrine cells may provide the ultimate method for intracerebral drug delivery. Transplanted secretory cells might provide all of the advantages of biological, as compared to mechanical, systems. They are self-replenishing, self-repairing "devices" which have the potential ability to deliver natural hormones, peptides, or secretory products directly to the desired site of action. Pappas and Sagen have exploited the enkephalin-secreting properties of adrenal chromaffin cells in animal models of pain.

Enkephalin is an endogenous opiate-like compound with pain-inhibiting effects. When transplanted to the spinal cord or central grey, chromaffin cells were found to inhibit pain in animal models, apparently through secretion of opiate peptides. This technique provides an excellent model for study of the possibilities for application of this kind of technique. Issues that would have to be addressed include the question of whether the specificity of the secretory products of transplanted cells would be sufficient to achieve the desired effect without prohibitive side effects, and whether these methods are sufficiently efficacious to be employed in the severe chronic pain syndromes where they would be the most useful.

The areas of investigation presented in this workshop represent a cross-section of areas of possible application of tissue engineering to problems related to nervous system functioning. In some cases, application to human clinical problems can be foreseen in the near future, whereas in other examples, these data provide information which can be of long-term use in the understanding of neuronal growth, differentiation, and recovery from injury. Many of the potential applications of tissue engineering to neuronal functioning cannot yet be foreseen, and much of the necessary information will be derived from basic studies of neuronal differentiation and disorders of the nervous system. There is a constant temptation to apply techniques prior to understanding of the underlying problems and goals. Nevertheless, there are already several well understood areas of application which can provide an impetus for further developments.

Tissue Engineering, pages 249–256
© 1988 Alan R. Liss, Inc.

ENGINEERING THE REGENERATION OF SENSORY FIBERS INTO THE
SPINAL CORD OF ADULT MAMMALS WITH EMBRYONIC ASTROGLIA
COATED MILLIPORE IMPLANTS

M. Kliot, G.M. Smith, J. Siegal, S. Tyrrell, C. Doller and
J. Silver

Dept. of Neurosurgery, Columbia Presbyterian Medical Center,
N.Y.C., N.Y. and Dept. of Developmental Genetics, Case Western
Reserve Univ., Cleveland, Ohio.

ABSTRACT In adult mammals, crushed dorsal root sensory
fibers do not normally regenerate beyond the entry zone
and into the spinal cord to any significant extent.
Using a pennant-shaped Millipore implant coated with
embryonic spinal cord astrocytes, it has been possible
to reduce scar formation and promote the regeneration
of crushed sensory axons across the dorsal root entry
zone and into the grey matter of the adult spinal cord.

INTRODUCTION

Although injured axons regenerate robustly within the
peripheral nervous system (PNS), their regrowth within the
adult mammalian central nervous system (CNS) is far more
limited (1). This failure of the CNS to regenerate has been
attributed to factors, such as the development of a glial
scar, in the immediate environment of the injured axon (14).
However Millipore implants (2), coated with astrocytes harvested from neonatal brain or embryonic spinal cord, have
been shown to reduce scar formation and promote the growth of
lesioned adult callosal (17,19) and corticospinal axons (15).
Although these coated implants stimulated fiber growth through
the original lesion site, axonal growth beyond the implant
was negligible.
The limited ability of adult mammalian neurons to regenerate axons within the CNS is clearly demonstrated by dorsal
root sensory fibers that are crushed (9,10,13). These fibers
grow successfully beyond the crush site within the PNS. How-

ever, at the cord surface, most of these regenerating axons
either make a U-turn and grow back towards the periphery or
make stable presynaptic-like endings upon the reactive astrocytes within the dorsal root entry zone (DREZ) (9,21). Only
a few fibers penetrate the DREZ and extend into the spinal
grey matter which is seperated by less than a millimeter
from the DREZ (10).

The astrocytes of the DREZ undergo reactive changes
(gliosis), including hyperplasia and proliferation, in response to degenerating axons (21). These changes are thought
to establish a permanent barrier to the passage of regenerating sensory fibers (14). In addition, it has recently been
suggested that a protein component of adult myelin, made by
the oligodendroglia, is a potent inhibitor of axonal elongation (4). Therefore the white matter of the posterior
collumns, as well as the tract of Lissauer, may constitute
additional obstacles in the path of regenerating fibers.

In contrast, the crushed dorsal root fibers of newborn
rats have been shown to regenerate successfully into the cord
and form appropriate connections within their target regions(3).
In an attempt to restore this regenerative potential to the
adult mammal, we have employed specially designed Millipore
implants coated with embryonic spinal cord astrocytes. Such
transplants can promote the growth of adult dorsal root fibers
across the DREZ and into the spinal grey matter where many
form axon terminals with boutons. In addition, examples of
abortive regeneration illustrated by axons passing into but
ending abruptly in white matter support the proposed inhibitory action of adult myelin on axonal elongation.

METHODS

Anaesthetized adult male and female albino rats (250-
400 gm) underwent bilateral laminectomies to expose the caudal
part of the spinal cord and nerve roots of the cauda equinae.
The fifth lumbar root (L5) on one side was isolated and loosely
knotted with 6-0 silk suture for future identification. This
root was then crushed, exerting maximal force three times for
ten seconds each, approximately 2-3 mm from its DREZ. In
some animals the adjacent one or two roots rostral and caudal
to the crushed root were cut.

Animals were divided into several experimental groups.
A control group was simply closed and allowed to recover. The
other groups were implanted with a Millipore structure in the
shape of a pennant. As shown in Fig. 1A, the flag portion

was inserted along the medial aspect of the L5 DREZ with the
pole portion lying just medial and adjacent to the nerve root.
One group of animals received "naked" Millipore pennants that
had not been coated with any cells. Another group received
implants coated with a purified population of astrocytes
isolated from the spinal cords of embryonic day 18 old rat
fetuses using the method described by Smith and Silver (20).
GFAP staining revealed ninety-five percent purity and almost
complete coverage of the implant surface with astrocytes
(Fig. 1B). Animals recovered for 3 to 4 weeks and they were
then reanaesthetized and the suture tagged L5 root re-isolated.
This root was cut 5 to 6 mm distil to its DREZ and a highly
concentrated solution of HRP was applied to its proximal end
for 1 to 2 hours. After 24 hours, the animals were perfused,
the L5 root-cord removed en bloc and vibratome sectionned,
and finally processed for HRP histochemistry using DAB with
Cobalt Cloride intensification.

RESULTS

    Animals with only the L5 root crushed and no Millipore
implant show substantial growth of fibers past the crush site.
At the DREZ axons either reverse direction and grow back
towards the periphery or enter the fringe of the cord where
they travel only for a short distance before ending abruptly
in a sterile club. Only rarely are fibers seen to proceed
past the fringe region and enter the dorsal horn grey matter.
    Animals with naked implants show a significant amount of
scar tissue at the interface between Millipore and spinal cord.
In several animals a severe reaction developed associated
with an intense infiltration of phagocytic cells and areas of
recurrent hemorrhage and cavitation in the surrounding spinal
cord. These animals show reduced ingrowth of axons into the
DREZ with none extending into the dorsal horn.
    Coating the implant with embryonic spinal cord astrocytes
greatly reduced the injury response and allowed the implant
to become well integrated with the parenchyma of the spinal
cord. Of particular interest is the presence of HRP-labeled
axons that penetrate the cord surface and in some cases
arborize extensively within the dorsal horn. Camera lucida
reconstructions at multiple levels of our best animal reveal
the vast majority of axons penetrating the spinal cord immediately adjacent to either the flag or pole portion of the
implant (Fig. 2). Along this interface of implant with cord
the tissue appears to be more loosely organized and infiltrated

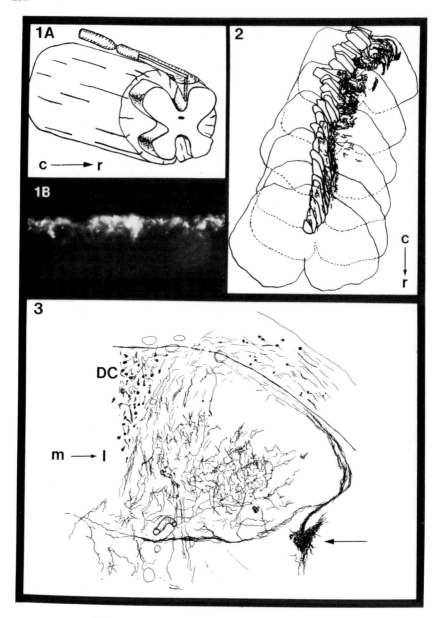

Figure 1. (A) A schematic representation of the 5th lumbar root and the spinal cord showing the placement of the pennant shaped implant. (B) A fluorescent photomicrograph showing GFAP staining of the astrocytes coating the implant.

with large numbers of phagocytic cells. Many of these labeled
fibers enter the dorsal horn and form normal appearing terminals (Fig. 3). Of note, however, is a group of fibers that
course laterally and superficially arround the dorsal horn and
terminate in a variety of bizarrely shaped clusters of densely
packed terminal boutons (arrow in Fig. 3). To our knowledge
such structures have not been described before (11,12,18).
Other fibers enter the myelinated territory of the dorsal
columns but extend only for short distances before ending in
either tight spirals or sterile clubs (Fig. 3:DC).

DISCUSSION

We have shown that controlling scar formation and providing crushed dorsal root fibers with an embryonic environment,
in which to bypass the intervening white matter, markedly
enhances regeneration across the DREZ and into the dorsal horn.
The close association of the HRP-labeled afferents with the
implant surface and the presence of unique terminal arbor malformations are strong evidence that these fibers actually
regenerated. Based on a comparison with our control animals,
we do not think sparing and/or sprouting of axons that would
be present regardless of the implant can account for these
results.

One of our most striking findings is the capacity of
regenerating fibers to arborize extensively over large and
grossly appropriate regions of the dorsal horn. This remar-
kable capacity for reinnervation is supportive of work suggesting ongoing as well as lesion induced synaptic plasticity
in the adult mammalian spinal cord (5,7,8).

Not all fibers arborize after penetrating the DREZ.
Many fibers freely enter myelinated territories. However,
once within them axons grow only for short distances,
often in tight spirals, and usually end in sterile clubs
without boutons. The in vitro studies of Schwab and

Figure 2. A three dimensional camera lucida reconstruction
in an animal with an embryonic astrocyte coated implant. Note
the close association of ingrowing HRP-labeled fibers (drawn
in black) with the implant surface.

Figure 3. A higher magnification camera lucida drawing of a
single caudal section shown in Fig. 2 (see text).

(c) caudal, (r) rostral, (m) medial, (l) lateral, (DC) dorsal
columns.

colleagues (4,16) provide evidence for the inhibition of axonal growth by adult white matter and our in vivo studies support this hypothesis.

Embryonic astrocytes reduce the ammount of scarring that occurs at the DREZ secondary to placement of the implant. Phagocytic activity is prominent in regions immediately adjacent to the astrocyte coated implant where axons penetrate the cord surface. It is possible that embryonic astrocytes initiate a cascade of cellular interactions that controls the injury response and promotes the growth of dorsal root fibers over widespread areas. Recently interactions between astrocytes and cells of the immune system have been demonstrated (6).

Our findings demonstrate the first method of reducing scar formation and engineering the regeneration of dorsal root afferents into appropriate regions of the adult mammalian spinal cord. This study provides hope that similar strategies employing embryonic astroglia may eventually lead to the repair of damaged intrinsic CNS pathways in animals and perhaps one day in humans.

ACKNOWLEDGEMENTS

We would like to thank Stacey Horn for excellent technical assistance. This work was supported by funds from the Daniel Heumann Fund for Spinal Cord Research, the Brumagin Memorial Fund of the Case Alumni Association, the American Paralysis Association, and the National Eye Institute of the NIH (#EY05952).

REFERENCES

1. Cajal SRY (1928). "Degeneration and Regeneration in the Nervous System." New York: Haffner.
2. Campbell JB, Bassett CAL, Husby J, Noback CR (1957). Regeneration of adult mammalian spinal cord. Science 126: 929.
3. Carlstedt T, Dalsgaard C-J, Molander C (1987). Regrowth of lesioned dorsal root nerve fibers into the spinal cord of neonatal rats. Neurosci Letts 74: 14.
4. Caroni P, Schwab ME (1987). Proteins of rat CNS white matter and of myelin forming oligodendrocytes are potent inhibitors of neurite outgrowth and of cell locomotion. Soc Neurosci Abstr 13: 1040.

5. Devor M, Wall PD (1981). Plasticity in the spinal cord sensory map following peripheral nerve injury in rats. J Neurosci 1: 679.

6. Fierz W, Fontana A (1986). The role of astrocytes in the interaction between the immune and nervous system. In Federoff S, Vernadakis A (eds): "ASTROCYTES Cell Biology and Pathology" Vol 3, Orlando: Academic Press, p 203.

7 Goldberger ME (1974). Recovery of movement after CNS lesions in monkeys. In Stein D (ed): "Recovery of Function After Neural Lesions," New York: Academic Press, p 265.

8. Liu CN, Chambers WW (1958). Intraspinal sprouting of dorsal root axons. Arch Neurol Psychit 79: 46.

9. Liuzzi FJ, Lasek RJ (1987). Astrocytes block axonal regeneration in mammals by activating the physiological stop pathway. Science 237: 642.

10. Liuzzi FJ, Lasek RJ (1987). Some dorsal root axons regenerate into the adult rat spinal cord. An HRP study. Soc Neurosci Abstr 13: 395.

11. Mawe GM, Bresnahan JC, Beattie MS (1986). A light and electron microscopic analysis of the sacral parasympathetic nucleus after labeling primary afferent and efferent elements with HRP. J Comp Neurol 250: 33.

12. Morgan C, Nadelhaft I, de Groot WC (1981). The distribution of visceral primary afferents from the pelvic nerve to Lissauer's tract and the spinal grey matter and its relationship to the sacral parasympathetic nucleus. J Comp Neurol 201: 415.

13. Perkins S, Carlstedt T, Mizuno K, Aguayo AJ (1980). Failure of regenerating dorsal root axons to regrow into the spinal cord. Can J Neurol Sci 7: 323.

14. Reier PJ, Stensaas LJ, Guth L (1983). The astrocytic scar as an impediment to regeneration in the central nervous system. In Kao CC, Bunge RP, Reier PJ (eds): "Spinal Cord Reconstruction," New York: Raven Press, p 163.

15. Schreyer DJ, Jones EG (1987). Growth of corticospinal axons on prosthetic substrates introduced into the spinal cord of neonatal rats. Dev Brain Res 35: 291.

16. Schwab ME, Theonen H (1985). Dissociated neurons regenerate into sciatic but not optic nerve explants in culture irrespective of neurotrophic factors. J Neurosci 5: 2415.

17. Silver J, Ogawa M (1983). Postnatally induced formation of the corpus callosum in acallosal mice on glia-coated cellulose bridges. Science 220: 1067.
18. Smith C (1983). The development and postnatal organization of primary afferent projections to the rat thoracic spinal cord. J Comp Neurol 220: 29.
19. Smith GM, Miller RH, Silver J (1986). Changing role of forebrain astrocytes during development, regenerative failure, and induced regeneration upon transplantation. J Comp Neurol 251: 23.
20. Smith GM, Silver J (1988). Transplantation of immature and mature astrocytes and their effect on scar formation in the lesioned CNS. Prog in Brain Res (submitted).
21. Stensaas LJ, Partlow LM, Burgess PR, Horch KW (1987). Inhibition of regeneration: The ultrastructure of reactive astrocytes and abortive axon terminals in the transition zone of the dorsal root. In Seil FJ, Herbert E, Carlson BM (eds): "Neural Regeneration," Amsterdam, Elsevier, p 457.

Tissue Engineering, pages 257–262
© 1988 Alan R. Liss, Inc.

THE ROLE OF BIOMATERIALS IN PERIPHERAL NERVE REGENERATION

P. Aebischer

Artificial Organ Laboratory, Brown University, Providence, Rhode Island 02912

ABSTRACT   The mechanisms underlying peripheral nervous system regeneration are poorly understood. As a consequence, no conceptual background to plan interventions exists, and recovery following peripheral nerve injury remains highly variable and incomplete. Synthetic guidance channels can serve effectively as tools to study the regeneration process and may be utilized clinically in the repair of injured nerves. Our work focuses on identifying specific physical and electrical properties of guidance channels which optimize regeneration and provide insight into the regeneration process. Three guidance channel characteristics that influence peripheral nerve regeneration have been identified; permeability characteristics, microgeometry and electrical activity.

Adult mammalian peripheral nervous system (PNS) axons are capable of regeneration following transection. Sprouting axons from the proximal nerve stump traverse the injury site, enter the distal nerve stump and make new connections with target organs. Current surgical techniques allow surgeons to realign nerve ends if the nerve defect is not too extensive. Nerve realignment increases the probability that extending axons will encounter an appropriate distal neural pathway, yet the incidence of recovery is highly variable and the return of function is never complete. Surgical advances in this area seem to have reached an impasse, since biological rather than technical factors now limit the quality of regeneration and functional recovery. Nerve guidance channels facilitate the reconnection of severed nerves in experimental animals and show promise for the repair of injured human nerves, although the material of choice has not yet been identified. Guidance channels are also valuable tools for elucidating the cellular and molecular processes which underlie nerve regeneration.

Advances in the synthesis of biocompatible polymers have provided scientists with a variety of biomaterials which may serve as nerve guidance channels. Biological materials such as arteries (1) and veins (2), and synthetic materials including silicone elastomer (3),polyethylene (48), polyvinyl chloride (5), acrylic copolymer (6) and bioresorbable polyesters (7) have been used experimentally. While much attention has been given to the biological events controlling the nerve regeneration process, little attention has been focused on the effect the guidance channel's physical properties exert on the regeneration process. In our approach, the nerve guidance channel is viewed as a material which interacts with the regeneration process by influencing the cellular and metabolic aspects of the regenerating environment.

The guidance channel's transport properties may influence the regeneration process by modulating solute exchange between the regenerating and extra-channel environments. Using the mouse sciatic nerve model, we have observed that permselective channels with a Mw cut-off of 50,000 daltons allowed the regeneration of nerves which more closely resemble the normal sciatic nerve (i.e. fine epineurium, high number of myelinated axons) than channels impermeable (silicone, polyethylene) or freely permeable (expanded polytetra-fluoroethylene) to watery solutes (8). We have also observed that in contrast to blind-ended impermeable tubes, blind-ended permselective tubes support extensive regeneration in the absence of a distal nerve stump (9). These experiments suggest that controlled exchange across the guidance channel wall enhances the formation of an optimal regenerating environment in addition to providing axonal guidance, preventing scar tissue invasion, minimizing the escape of growth factors released by the nerve stumps and preventing the release of nerve antigenic factors. Most studies dealing with peripheral nerve regeneration use guidance channels which have little or no permeability to watery solutes; silicone elastomer (3), polyethylene (4) and polyvinylchloride (5). These materials were selected to isolate the regenerative environment from the surrounding tissues so that only cells or fluids within the chamber would influence regeneration, but they eliminate the effect of numerous external factors that could also affect the process.The use of permselective materials with a range of Mw cut-offs allows both intra- and extra-channel factors to be studied and provides enhanced metabolic support.

In attempts to elucidate the biological mechanisms regulating axonal growth following peripheral nerve transection, several research groups have investigated a variety of distal tissue inserts in straight or Y-shaped guidance channels. A lack of regeneration when the distal end of impermeable tubes was capped or left open and the distal nerve stump avulsed has been reported by several groups (10). Placing a cut tendon at the distal end of a silicone tube leads, in the best case, to the formation of a small granulation tissue cable with minimal axonal elongation (10). A preferential or exclusive tendency for the regenerating nerve to grow toward the distal nerve insert versus the tendon insert or an empty branch has been reported with silicone Y-shaped tubes (11). Using materials of biological origin, Weiss and Taylor reported contadictory results (12). Using the iliac bifurcation of the rat abdominal aorta, they observed similar degrees of growth toward a distal nerve as compared to distal tendons or blind-ended branches. The use of a biological tissue may have confused the experiment by providing non-neurally produced growth factors which influenced the regeneration process. In preliminary experiments we have observed extensive regeneration from a proximal nerve stump placed in a capped, permselective tube. A parallel experiment using capped silicone tubes showed no regeneration. Using semi-permeable Y-shaped tubes, we have observed the regeneration of myelinated axon-containing cables toward both distally placed nerve stumps and tendons. These observations suggest that growth factors possibly secreted by the wound healing process outside of the guidance channel may diffuse within the regenerating environment and favor the regeneration process. The use of a synthetic, permselective tube offers a way to separate these factors.

A second physical characteristics of the guidance channel we are adressing is the luminal surface morphology or microgeometry. Nerves regenerated in nerve guidance channels are usually centrally located in the guidance channel and surrounded by an acellular gel. The nerve cables do not touch the channel's inner surface. We have observed, however, that the use of a channel with a relatively rough microfibrillar inner surface resulted in regenerated nerve fascicles lacking an epineurium and spread in a loose connective tissue which entirely filled the channel lumen (13). Rough synthetic surfaces are known to induce a stronger tissue reaction than smooth surfaces in both the subcutaneous and intramuscular locations. We hypothetize that the morphological arrangement of the regenerated nervous tissue is affected by the interaction of the channel material with the nervous tissue. The cellular/polymer contact occurs only at the level where the native nerve stumps contact the guidance channel. It is possible that the epineurial cells instead of growing on the periphery of the fibrin cable, are trapped in the voids of the channel wall and therefore cannot delineate a single cable for cellular migration. By determining how the surface geometry and channel permerability effect the early events of regeneration we hope to better understand the process of initial cable formation in peripheral nerve regeneration.

The electrical properties of guidance channels may also influence the nerve regeneration process. Since in vitro neurite outgrowth has been shown to be promoted and directed by electrical activity (14), it is conceivable that guidance channels displaying electrical activities could favor nerve regeneration in vivo, as already achieved by the application of pulsed electromagnetic fields (15) or D.C. stimulation (16). These latter methods entail the use of large electromagnets and power supplies with implanted wiring. Guidance channels composed of charged polymers would preclude the need for an external power source or electrical circuitry. Electrets are dielectric materials exhibiting quasi-permanent electrical charges, the term quasi-permanent meaning that the time constant characteristics for the charge decay are much longer than the time periods over which studies are performed. The electret charges may consist of 1) "real" charges, usually due to layers of trapped positive or negative carriers positioned at or near the two surfaces of the dielectric material, or 2) "true" polarization, i.e. a permanent alignment of dipoles. An electret may also combine "real" charges and "true" polarization. Electrets not covered by metal electrodes usually produce an external electrostatic field and the corresponding analogy with a magnet is often used to define them. Under this broad definition, piezoelectric materials also belong to the electret category. Current practice tends to use the term "electret" to indicate those materials in which the electrical charge is stored predominantly as trapped surface charge, whereas piezoelectric materials are characterized by the need for some form of mechanical deformation to generate a transient electric field.

Many synthetic and biological materials can be prepared as electrets, usually by "poling" them in a high intensity electrical field. Polytetrafluoroethylene (PTFE or teflon) and polyvinylidene fluoride (PVDF or PVF2) have been investigated for a number of different biomedical applications. They display good in vivo biocompatibility and have been reported to enhance osteogenesis (17). Our preliminary experiments indicate that poled PVDF nerve guidance channels enhance peripheral nerve regeneration as compared to unpoled PVDF tubes (18). Piezoelectric PVDF channels provide a useful model for the anlaysis of the effect of charge sign (positively versus negatively poled PVDF tubes), the magnitude of charge generation, and the relative contribution of each nerve stump on the regeneration process by poling only part of the tube.

In the past, synthetic guidance channels have been considered as inert conduits providing axonal guidance, maintaining growth factors, and preventing scar tissue invasion. Little or no emphasis has been placed on the relationship between the physico-chemical properties of the channel and the outcome of regeneration. These relationships must be elucidated since, experimentally, the morphological and functional results of regeneration compare poorly with the normal situation. Identification of the physical and electrical characteristics of nerve guidance channels which optimize peripheral nerve regeneration may lead to a better understanding of the regeneration process and allow the design of optimal guidance channels for use in the clinical repair of severed or avulsed nerves.

REFERENCES

1. Weiss P (1944). The technology of nerve regeneration; a review. Sutureless tubulation and related methods of nerve repair. J Neurosurg 1: 400.
2. Chiu TWD, Janecka I, Krizek TJ, Wolff M, Lovelace RE (1982). Autogenous vein graft as a conduit for nerve regeneration. Surgery 91: 226.
3. Lundborg G, Gelberman RH, Longo FM, Powell HC, Varon S (1982). In vivo regeneration of cut nerves encased in silicone tubes: growth across a six-millimeter gap. J Neuropathol Exp Neurol 41: 412.
4. Da Silva CF, Madison R, Greatorex D, Dikkes P, Sidman RL (1985). Quantative effects of a laminin-containing gel, collagen, or empty polyethylene tube on peripheral nerve regneration in vivo. Soc. Neurosci Abstr 11: 1253.
5. Scaravilli F (1984). Regeneration of the perineurium across a surgically induced gap in a nerve encased in a plastic tube. J Anat 139: 411.
6. Uzman BG, Villegas GM (1983). Mouse sciatic nerve regeneration through semipermeable tubes: a quantitative model. J Neurosci Res 9: 325.
7. Molander H, Olsson Y, Engkvist O, Bowald S, Eriksson I (1982). Regeneration of peripheral nerve through a polyglactin tube. Muscle and Nerve. 5: 54.
8. Aebischer P, Valentini RF, Winn SR, Kunz SK, Galletti PM (1986). Are guidance channel composition and structure important factors in peripheral nerve regeneration? Soc. Neurosci. Abstr. 12: 699.
9. Aebischer P, Guénard V, Winn SR, Valentini RF, Galletti PM (1988). Blind-ended semi-permeable guidance channels support peripheral nerve regeneration in the absence of a distal nerve stump. Brain Res (in press)

10. Lundborg G, Dahlin LH, Danielsen N, Gelberman RH, Longo FM, Varon S (1982). Nerve regeneration in silicone chambers; influence of gap length and of distal stump components. Exp Neurol 76: 361.
11. Politis MJ (1985). Specificity in mammalian peripheral nerve regeneration at the level of the nerve trunk. Brain Res 328: 271.
12. Weiss P, Taylor AC (1944). Further experimental evidence against "Neurotropism" in nerve regeneration. J Exp Zool 95: 233.
13. Aebischer P, Guénard V, Winn SR, Galletti PM (1988). Non-porous versus expanded polytetrafluoroethylene (PTFE) tubes as peripheral nerve guidance channels. ASAIO Abstr 17: 70.
14. Jaffe LF, Poo MM (1979). Neurites grow faster toward the cathode than the anode in a steady field. J Exp Zool 209: 115.
15. Ito H, Bassett CAL (1983). Effect of weak, pulsing electromagnetic fields on neural regeneration in the rat. Clin Ortho Rel Res 181: 283.
16. Kerns JM, Freeman JA (1986). D.C. electrical fields promote regeneration in the rat sciatic nerve after axotomy. Soc Neurosci Abstr 12: 13.
17. Fukada E (1981). Piezoelectricity of bone and osteogenesis by piezoelectric films. In: Becker RO, Thomas CC (eds) Mechansims of growth control, Thomas, Springfiled, p. 192.
18. Aebischer P, Valentini RF, Dario P, Domenici C, Galletti PM (1987). Piezoelectric guidance channels enhance regeneration in the mouse sciatic nerve after axotomy. Brain Res. 436: 165.

Tissue Engineering, pages 263–267
© 1988 Alan R. Liss, Inc.

RECIPROCAL CONTROL OF NEUROBLASTOMA NEURITE OUTGROWTH BY PROTEASE NEXIN-1 AND THROMBIN[1]

Dennis D. Cunningham and David Gurwitz[2]

Department of Microbiology and Molecular Genetics
College of Medicine
University of California, Irvine, CA 92717

ABSTRACT An understanding of the formation and regulation of neuronal networks will depend on identifying factors which regulate both the outgrowth and retraction of neurites and learning how they do this. The control of neurite extension is likely controlled by a complex series of positive and negative signals. Here we summarize: (a) studies on the ability of a protease inhibitor (glial-derived neurite promoting factor/protease nexin-1) to stimulate neurite outgrowth in cultured neuroblastoma cells; (b) evidence that its neurite outgrowth activity is due to inhibition of thrombin; and (c) studies on the ability of thrombin to retract neurites in cultured neuroblastoma cells.

INTRODUCTION

The regulation of neurite outgrowth from neuronal cells has been studied intensively because of its central role in development of the nervous system and its potential importance in neural regeneration. It is clear that an understanding of this regulation will be facilitated by knowledge of the factors which control neurite outgrowth and an understanding of how this cellular regulation is mediated. This is particularly clear from the classic studies on nerve growth factor (1). A number of neurotrophic factors have been identified, and with the realization that certain factors affect only certain kinds of neurons, it appears that the number of factors involved in this complex regulation is probably large.

---

[1] This work was supported by National Institutes of Health Grant CA12306.
[2] Recipient of a Chaim Weizmann Postdoctoral Fellowship.

Although much emphasis has been placed on identification of factors that stimulate neurite outgrowth, there is little information on physiological agents which can modulate or reverse this process. It is important to elucidate mechanisms of neurite retraction since axon elimination occurs during development (2) and since regulation of neurite outgrowth probably involves an intricate coordination of positive and negative signals.

Based on previous studies which identified a glial-derived neurite promoting factor with protease inhibitory activity (3), we recently showed that thrombin can bring about retraction of neurites from neuroblastoma cells and that these two factors can reciprocally control neurite outgrowth (4).

RESULTS

Recent studies have shown that the glial-derived neurite promoting factor identified by Monard and colleagues is identical to protease nexin-1 (PN-1) (5,6). PN-1 is a 43 kDA protease inhibitor that is secreted by cells and which complexes certain serine proteases including thrombin (7); the complexes then bind back to the cells and are internalized and degraded (8). Figure 1 (panel A) shows the ability of purified PN-1 to stimulate neurite outgrowth from cultured mouse neuroblastoma cells (clone Nb2a). As shown, added PN-1 increased the fraction of cells with neurites longer than one cell diameter from about 12 percent to over 60 percent. Added thrombin blocks this effect (Figure 1, panel A) (4). The stimulation of neurite outgrowth by PN-1 appears to be due to its inhibition of thrombin since hirudin, a potent thrombin inhibitor that is structurally different from PN-1, also stimulates neurite outgrowth from neuroblastoma cells (3,4).

In addition to blocking neurite outgrowth brought about by PN-1, added thrombin also produces neurite retraction after neurites are extended by treatment with PN-1 (Figure 1, panel B). The retraction of neurites by thrombin requires its proteolytic activity. The ability of thrombin to retract neurites, however, appears not to be a result of generalized proteolysis that might simply detach neurites from the culture dish since much higher concentrations of urokinase, plasmin and trypsin do not significantly bring about retraction of neurites (4). When neurites are extended by placing neuroblastoma cells in serum-free medium (9), thrombin blocks and reverses this process with a half-maximal potency of 50 pM. Together, these results indicate that thrombin can specifically modulate and reverse neurite outgrowth (4).

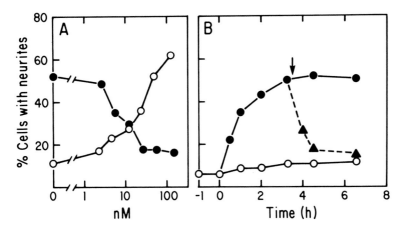

FIGURE 1. Induction of neurite outgrowth by PN-1 and its inhibition and reversal by thrombin. Panel A: The medium on neuroblastoma cells was changed from growth medium containing 10% serum to medium containing 0.8% serum. Then, the indicated concentrations of PN-1 (o) or 46 nM PN-1 along with the indicated concentrations of thrombin (●) were added and 4 h later the cells were scored for neurite outgrowth. Panel B: The medium on neuroblastoma cells was changed to medium containing 0.8% serum at time zero and the percentage of cells with neurites was scored in cultures without additions (o). In parallel cultures 46 nM PN-1 was added at time zero (●); at 3.5 h (arrow) thrombin (50 nM) was added to some of these cultures (▲). Reprinted with permission from (4).

DISCUSSION

The studies described above employed purified PN-1 and thrombin and were conducted on cloned neuroblastoma cells, permitting the conclusion that the proteins acted directly on these cells and not via glial cells. An important question is whether the reciprocal regulation of neurite outgrowth by PN-1 and thrombin also occurs with normal neurons. It is noteworthy that Monard and colleagues recently reported that the glial-derived neurite promoting factor/PN-1 also stimulates neurite outgrowth from primary neuronal cultures (10). Since the neurite outgrowth activity of this factor appears due to its inhibition of thrombin (3,4), it seems likely that thrombin brings about neurite retraction in certain neuronal populations, although to date this activity has been studied only in neuroblastoma cells. It is noteworthy that PN-1 binds to the extracellular matrix of cultured fibroblasts (11) and that this interaction accelerates its inhibition of thrombin (12) and blocks its ability to form complexes with urokinase and plasmin (13). Recent studies have shown that cultured neuroblastoma and glioma cells also accelerate the inactivation of thrombin by PN-1 and block the ability of PN-1 to inhibit urokinase and plasmin[3]. This regulation of PN-1 would increase its neurite promoting activity.

A major question regarding the role of thrombin in the regulation of neurite outgrowth in vivo is the conditions under which neurons would be exposed to this protease. The known conditions that lead to the processing of plasma prothrombin to thrombin involve injury or trauma. Although this could lead to neurite retraction that often occurs after injury to the brain, it is noteworthy that recent preliminary results[4] indicate that brain and cultured neuroblastoma cells contain prothrombin mRNA. If indeed neuronal cells synthesize prothrombin and process it to thrombin, this could provide, along with the glial-derive neurite promoting factor/PN-1, a localized mechanism for reciprocally regulating neurite outgrowth in vivo.

In future studies it will be important to address the controls underlying the synthesis and secretion of these and other factors that control neurite outgrowth. This could provide valuable information about the regulation of neurogenesis, and it could also provide clues about possible therapies to be used in neurodegenerative diseases or neural injury. For example, an understanding how the synthesis and secretion of neurotrophic factors is locally controlled could lead to a controlled delivery of these factors either by endogenous cells of the nervous system or implanted fetal cells.

REFERENCES

1. Levi-Montalcini R (1987). The nerve growth factor 35 years later. Science 237:1154.

2. Schreyer DJ, Jones EG (1988). Axon elimination in the developing corticospinal tract of the rat. Dev Brain Res 38:103.

3. Guenther J, Hanspeter NK, Monard D (1985). A glia-derived neurite promoting factor with protease inhibitory activity. EMBO J 4:1963.

4. Gurwitz D, Cunningham, D (1988). Thrombin modulates and reverses neuroblastoma neurite outgrowth. Proc Natl Acad Sci USA (in press).

---

[3] Wagner S, Lau A, Cunningham, D (unpublished results).
[4] Wagner S, Isackson P, Cunningham D (unpublished results).

5. Gloor S, Odink K, Guenther J, Nick H, Monard D (1986). A glia-derived neurite promoting factor with protease inhibitory activity belongs to the protease nexins. Cell 47:687.

6. McGrogan M, Kennedy J, Li M, Hsu C, Scott R, Simonsen C, Baker J. (1988). Molecular cloning and expression of two forms of human protease nexin 1. Bio/Technology 6:172.

7. Baker J, Low D, Simmer R, Cunningham D. (1980). Protease nexin: a cellular component that links thrombin and plasminogen activator and mediates their binding to cells. Cell 21:37.

8. Low D, Baker J, Koonce W, Cunningham D. (1981). Release protease nexin regulates cellular binding, internalization and degradation of serine proteases. Proc Natl Acad Sci USA 78:2340.

9. Bottenstein J, Sato G. (1979). Growth of a rat neuroblastoma cell line in serum-free supplemented medium. Proc Natl Acad Sci USA 76:514.
10. Monard D, Gloor S, Sommer J, Hanspeter N. (1987). Biological relevance of a glia-derived protease inhibitor. J. Neurochem. 48 (Supplement):S17.

11. Farrell D, Wagner S, Yuan R, Cunningham D. (1988). Localization of protease nexin-1 on the fibroblast extracellular matrix. J Cell Physiol 134:179.

12. Farrell D, Cunningham D. (1986). Human fibroblasts accelerate the inhibition of thrombin by protease nexin. Proc Natl Acad Sci USA 83:6858.

13. Wagner S, Lau A, Cunningham D. (submitted for publication).

Tissue Engineering, pages 269–271
© 1988 Alan R. Liss, Inc.

Summary/Discussion

Restoration and Maintenance of Neurological Function

Michel Kliot

This session identified conditions, both physical and biochemical in the immediate environment of neuronal processes, that can influence their growth and maintenance. Patrick Aebischer identified several important properties of synthetic polymer tubes that can significantly affect the regeneration of mammalian peripheral nerves, and Dennis D. Cunningham defined several biochemical factors that can control the extension and retraction of neuronal processes.

Aebischer has constructed acrylic polymer tubes which serve as conduits for regenerating peripheral nerves. By selectively altering the permeability, electrical charge and microgeometry of these tubes, he has been able to study the effects that each of these properties has on axonal regeneration. These tubes consisted of an inner and outer membrane with an intervening trabecular meshwork.

The outer membrane was sufficiently porous to allow extensive penetration of capillaries up to, but not beyond, the permselective inner membrane. Excellent growth of fibers occurred through the tubes when the permeability of the inner membrane was set at 50,000 daltons. Many of these fibers became myelinated. The number of total fibers, as well as the number of myelinated fibers, was substantially greater in these permselective tubes as compared to impermeable polyethylene or silicone tubes. In addition, closing off the distal end of these permselective tubes did not abolish the ingrowth of nerve fibers as was found to occur with impermeable tubes. However, capping the distal end of the tube did reduce the average diameter of the regenerating axons. Of particular interest was the close association of macrophages lining the external surface of the inner wall of the tube with growing nerve fibers. Dr. Aebischer commented on the fact that these cells have been shown to secrete several factors (i.e. nerve growth factor, fibroblast growth factor and apolipoprotein E) which promote axonal growth.

Aebischer also demonstrated how electrical activity can influence the growth of axons. He employed tubes made of a piezoelectric polymer (polyvinyl fluoride) whose surface is charged by placing it in a strong electric field. Such electrical poling of the tubes increased the number of myelinated axons growing within the tubes. Finally, altering the texture of tube surface, without changing its permeability, was also shown to affect the amount of axonal ingrowth.

These tubes are currently being used to repair peripheral nerve injuries in primates. Preliminary studies suggest that improved functional recovery and faster nerve conduction occur when these charged permselective tubes are employed as compared to impermeable electrically neutral tubes.

Kliot reported that Millipore (R) filters impregnated with embryonic astrocytes and placed appropriately will induce regeneration of dorsal root fibers into the gray matter of the adult mammalian spinal cord. This is an important finding as the first significant report showing CNS regeneration.

Cunningham discussed several biological molecules that can control either the extension or the retraction of neuronal processes from a human neuroblastoma cell line grown *in vitro*.

Protease Nexin-1 (i.e., PN-1) is a 43,000-dalton molecule secreted by a number of different cell types. Although it was originally isolated from fibroblasts, more recently it was found to be identical to a molecule known to be secreted by astrocytes of the central nervous system.

PN-1 was shown by Cunningham to significantly stimulate neurite outgrowth of neuroblastoma cells as well as primary cultures of neurons. This stimulating effect can be blocked by thrombin. PN-1 binds thrombin as well as urokinase, plasmin, and components of the extracellular matrix. In addition, PN-1 binds to receptors on the surface of cells which secrete it. This binding leads to the rapid internalization of PN-1 and provides a feedback mechanism by which the molecule could regulate its own production and secretion.

Thrombin itself was shown to cause the retraction of neurites from neuroblastoma cells. Hirutin, a specific inhibitor of thrombin, had an opposing effect. Interestingly, the cultivation of neuroblastoma cells in serum-free medium led to enhanced neurite outgrowth. The addition of thrombin to serum-free medium caused the retraction of neurites already present and also blocked the formation of new neurites. Thrombin was shown to be effective at extremely low concentrations and required the presence of its proteolytic activity.

Finally, the question of whether these biomolecules normally regulate the extension and retraction of neuronal processes was addressed. PN-1 is secreted by astrocytes of the CNS. Thrombin accumulates in areas of injury, where breakdown of the blood-brain barrier occurs, and therefore could contribute to nerve tissue damage by promoting the retraction of neuronal processes.

In addition, the presence of prothrombin within the mammalian CNS has been confirmed using a cDNA probe. Thus, there is evidence to suggest that these biomolecules are normally present within the nervous system and may be playing a role in regulating the growth and maintenance of neuronal processes. A better understanding of their function may, therefore, lead to new strategies for reducing injury and promoting regeneration within the mammalian nervous system.

Tissue Engineering, pages 273–282
© 1988 Alan R. Liss, Inc.

INTRACEREBRAL NEUROTRANSMITTER REPLACEMENT THERAPY
FOR PARKINSON'S DISEASE

Saul S. Schwarz and William J. Freed*

Department of Neurosurgery, Naval Hospital Bethesda, and
*Preclinical Neurosciences Section,
National Institute of Mental Health,
Saint Elizabeths Hospital, Washington, D.C. 20032

ABSTRACT Adverse effects and eventual failure of L-
DOPA in the treatment of Parkinson's disease have
prompted an investigation into intracerebral methods of
neurotransmitter replacement. Various intracerebral
drug infusion and tissue transplantation experiments
have been performed, and these techniques are reviewed
so far as they may offer solutions to the problems of
current medical therapy.

INTRODUCTION

Recent research efforts in the field of brain tissue
transplantation are directed at exploring treatment
techniques applicable to human degenerative brain disorders.
Parkinson's disease, a progressive and devastating complex of
rigidity, tremor, loss of postural reflexes, slowing of
movement and late-onset dementia, affects roughly 200,000
people in this country and arises in 40,000 new cases each
year (1). A basic research paradigm for a spectrum of
neurodegenerative disorders of unknown etiology, Parkinson's
disease is thus in itself a major health care problem.

Current medical therapy is based upon the discovery that
brains of Parkinson's disease patients are profoundly
depleted of dopamine, a catecholamine neurotransmitter
produced in midbrain substantia nigra (SN) neurons and
transported up to neurons in the corpus striatum (2).
Degeneration and loss of these dopaminergic substantia nigra
neurons are consistently seen and are characteristic of this
disease. Treatment of patients with L-DOPA (L-dihydroxy-
phenylalanine), a precursor of dopamine which crosses the
blood-brain barrier, set a precedent in clinical neurology in
that it was conceived as a neurotransmitter replacement
therapy. Although originally a major breakthrough in the

1960s, L–DOPA given systemically is associated with several adverse acute and long–term effects which ultimately offset its benefit for a majority of patients (3). Dopamine receptor agonists such as bromocriptine and pergolide have been tried alone or as adjuvant therapy with L–DOPA; preliminary reports suggest these agents may smooth out the therapeutic effects of L–DOPA during the early disease phase (4). Long–term medical failure has prompted investigators to search for better ways of replacing the lost neurotransmitter.

REPLACEMENT THERAPY RESEARCH

Techniques explored in the laboratory have included direct infusion of dopamine into the striatum or adjacent ventricle, implantation of cell suspensions or tissue fragments of embryonic substantia nigra to these sites, and transplantation methods using other dopaminergic tissues, such as adrenal medulla or PC12 pheochromocytoma cells.

Each of the stumbling blocks in current systemic dopamine therapy offers a perspective for critiquing these various methods:

Acute Systemic Side–Effects of Dopamine Therapy

L–DOPA taken systemically produces such peripheral symptoms as nausea, palpitations and cardiac arrhythmias, and postural hypotension. The compounding of L–DOPA with carbidopa (alpha–methyldopa hydrazine), a decarboxylase inhibitor which does not cross the blood brain barrier, has reduced some of these peripheral dopamine affects while significantly lowering the dose of L–DOPA required to achieve therapeutic brain levels. Combination therapy (Sinemet) is currently the most common form of L–DOPA prescribed in this country. Preferably, this problem could be solved more directly by any method of local dopamine delivery to the striatum, restricted in its action by limited tissue diffusion and local uptake or degradation.

Hood et al. (5) reported intraventricular pump infusion of L–DOPA methyl ester (a water soluble dopamine precursor) in monkeys lesioned with MPTP (1–methyl–4–phenyl–1,2,3,6–tetrahydropyridine hydrochloride), a selective nigrostriatal neurotoxin which produces advanced Parkinson–like rigidity, bradykinesia and impaired feeding. Reversal of bradykinesia and feeding impairment was noted in short–term trials. Intraventricular infusion of dopaminergic agents may produce undesirable distant effects, such as agitation or psychosis, because of widespread diffusion through the brain. Ventricular infusion of L-DOPA in rats tends to also produce marked aggression (6).

Acute Onset Involuntary Movements of Dopamine Therapy

Either alone or coupled with carbidopa, L-DOPA may produce dyskinesia such as facial grimacing, tongue thrusting and lip smacking, dystonic posturing of the extremities or trunk, motor tics and a generalized bodily restlessness. This cluster of hyperkinetic responses is thought to represent dopamine toxicity in the striatum and mesolimbic system, and is related to peak fluctuations in blood levels. Most patients initially accept some toxic side-effects along with the salutary effect of L-DOPA, but the dyskinesias tend to become more severe and frequent with chronic therapy. A direct striatal pump infusion promises accurate and nonfluctuating local drug levels without pharmacological action or interference by other brain regions.

Using a unilateral nigrostriatal lesion model in the rat, Hargraves and Freed (7) have demonstrated attenuation of apomorphine-induced turning by continuous dopamine infusion through a cannula placed stereotactically into the corpus striatum. Adverse behavioral changes were not noted, and the drug infusions appeared to be as effective, at least in short-term, as tissue transplants in this experimental model.

The physiology of the nigrostriatal motor system is not, however, well understood. Would optimal design for human use require a continuous or pulsed infusion, or would moment-to-moment adjustments in dopamine release be necessary, modeled after hypothetical nigrostriatal feedback circuits(8)? Any method of dopamine delivery which mimics its natural synaptic delivery in the striatum would be optimal. To avoid the toxicity of oral L-DOPA therapy, chronically implanted dopamine infusion systems might require feedback sensitivity or patient control of flow rates.

Dose-Response Fluctuations of Dopamine Therapy--the "On-Off" Phenomenon

Chronic use of systemic L-DOPA leads to wide swings in drug efficacy in which the patient may have normal or even excessive mobility one moment and become suddenly, unpredictably frozen with rigidity and tremor the next. Over years, as neuronal loss in the substantia nigra continues, an "up-regulation" or increase in striatal post-synaptic receptors develops which creates a supersensitivity to dopamine agents (9,10). Additionally, PET scanning studies have demonstrated an impaired dopamine storage capacity in the chronically denervated striatum, particularly in those patients exhibiting the on–off phenomenon (11). This may account for the "wearing-off", or abrupt nadir effect of L–DOPA seen with long-term usage. At least some of the action of L–DOPA appears to be dependent upon uptake and release of converted dopamine by presynaptic nigrostriatal terminals, which continue to degenerate as the disease progresses (4). Occurring within five to eight years of L–DOPA treatment in over half the patients studied, severe episodic fluctuations may require extended in–hospital drug holidays and become as incapacitating as Parkinson's disease itself (3). This underscores the fact that dopamine therapy—whether administered systemically or by direct brain infusion—is a transmitter replacement and not a tissue replacement. Transplantation of dopaminergic tissue may potentially solve the problem of failed presynaptic dopamine storage, because the implant is self-replenishing.

The rat adrenal medulla demonstrates high concentrations of dopamine when transplanted to the lateral ventricle, adjacent to the striatum (12). Adrenal medulla grafts have been shown to reverse apomorphine–induced rotational behavior in the unilaterally lesioned rat. Bathed by spinal fluid, the survival rate for these ventricular grafts appears to be good, although the cells produce neurite–like processes without reinnervating the host brain. The graft effect may be related to delivery of dopamine into the surrounding striatum, as noted in fluorescence histochemical studies (13). The initial success with human intraventricular adrenal medulla transplantation for Parkinson's disease, as reported by Madroza et al. (14), appears to be bilateral, despite unilateral grafting. Although cerebrospinal fluid diffusion of dopamine into contralateral striatum has been postulated, it is unlikely that grafted tissue could achieve adequate ventricular levels of dopamine to affect the brain bilaterally (15). Long–term results from the growing number of human adrenal medulla transplantation trials may clarify mechanisms of transplant action. It remains to be seen whether a graft which functions as a "biological pump" without host reinnervation can solve the "on–off" problem noted with chronic L–DOPA therapy.

Direct striatal implants of adrenal medulla do not survive well (16, 17) nor do implants of dissociated adrenal medullary cell suspensions (18). This may account for the modest, transient success of stereotactic intrastriatal adrenal medullary tissue implants in patients with Parkinson's disease, reported by Backlund et al. (19). Striatal infusion of nerve growth factor may enhance survival of these direct striatal implants, and promote host reinnervation (20). Other strategies used to promote neurite ingrowth include injuring the cerebral cortex overlying the intrastriatal graft (21), and preparing an implant bed days or weeks prior to transplantation (22).

Catecholamine–secreting cell lines may provide an alternate and potentially infinite source of dopamine for transplantation. Cultured PC12 pheochromocytoma cells have been transplanted into the rat striatum with a moderate rate of long–term survival, some cellular sprouting with neurite–like processes, and no evidence of uncontrolled tumor growth (23). Nerve growth factor has been shown to further control growth and enhance neuronal differentiation in these cells (24).

The problem of denervation sensitivity and other chronic reactive changes in the striatum might only be solved by reinnervation of the host brain. Theoretically, donor tissue identical in cell population to the degenerated host tissue might extend neurites which "recognize" striatal neurons as appropriate targets for synapse formation (25). Homologous tissue has the added advantage of "bringing along for the ride" additional neurotransmitters, neuropeptides and growth factors as yet unappreciated in this disorder. For these reasons, embryonic SN is ultimately the most promising transplant tissue for human use.

Transplantation of embryonic SN to the lateral ventricle in rats yields excellent survival rates with dopaminergic innervation of the adjacent host striatum, as demonstrated by catecholamine fluorescence histochemistry (13) and in vivo monoamine electrochemistry (26). Using the unilateral nigra lesion paradigm, SN ventricular implants achieve a high reversal rate in apomorphine–induced rotational behavior (13,27). When implanted bilaterally in the ventricle of newborn rats, embryonic SN has been shown to protect mature rats from the akinesia, rigidity and impaired feeding produced by bilateral nigra lesions (28). Transplanted embryonic SN neurons demonstrate spontaneous electrical activity similar to normal SN neurons, and are inhibited by dopamine agonists and stimulated by dopamine antagonists (29). Spiroperidol autoradiography studies have shown that dopaminergic supersensitivity in the striatum is markedly reduced in areas reinnervated by embryonic SN grafts (30). Using tyrosine hydroxylase immunocytochemistry, synaptic contacts between grafted SN neurites and host striatal neurites have been demonstrated which are similar to those occurring naturally between nigral and striatal cells; a hyperinnervation pattern of striatal cells has also been observed (31). Thus embryonic SN grafts hold the greatest

promise for avoiding the debilitating side effects which now
limit L-DOPA usage.

Progression of Dementia and Non-Motor Symptoms

    As patient survival has improved on L-DOPA therapy there
has been an increasing recognition that dementia may be an
integral   component   of   Parkinson's   disease.   Postural
instability and depression also tend to progress despite
long-term L-DOPA treatment.   Although nigral degeneration is
most   prominent,   diffuse   neuronal   loss   occurs   in   other
pigmented nuclei of the brainstem, and alterations in brain
levels have been noted with several other transmitters
(norepinephrine,   serotonin,   gamma-aminobutyric   acid)   and
neuropeptides as well (32,33).   Diffuse cortical changes
typical of Alzheimer's disease are seen in many end-stage
Parkinson's   patients,   indicating   that   L-DOPA   therapy   is
directed at but one aspect of a diffuse degenerative process
(34).   Thus, any therapy which replaces lost transmitters or
tissue may improve some aspects of the disease, but might
leave other aspects untouched.
    One related limitation with brain tissue grafting is the
small   area   of   reinnervation   possible   with   a   tissue
transplant.   Despite   marked   reversal   of   lesion-induced
behavioral abnormalities, there is no evidence of synapse
formation or dopaminergic reinnervation beyond 0.5-2.0 mm
from the graft border (13,21,31).   Functional improvement may
be related to the intensity of regional reinnervation, and to
exceeding some crucial innervation "threshold" below which
behavioral   deficits   appear.   Still,   failure   to   correct
certain   deficits   in   the   lesioned   rat   model   with   SN
transplants has been related to a regional specialization of
function within the corpus striatum (35,36).   In an effort to
increase   the   area   of   reinnervation,   Dunnett   et   al.   have
performed intrastriatal implantation of SN cell suspensions
to multiple sites bilaterally in the rat.   This technique has
been shown to correct the rigidity, akinesia and sensorimotor
deficits, but not the eating and drinking deficits seen after
bilateral nigral lesions (36).
    It has been estimated that the zone of reinnervation of a
0.5-1mm tissue graft is restricted to a 1.5 mm radius (21).
The zone of effective dopamine diffusion by an intrastriatal
pump falls off by two orders of magnitude within a 6mm
radius, as measured with tritiated dopamine radiography (7).
Given the enormous size of the human striatum (roughly 250
times   larger   than   the   rat   striatum)   (37),   it   may   be
advantageous   to   work   out   a   regional   specialization   of

function in the human striatum, if any, to make intracerebral drug infusion or transplantation feasible for clinical use.

CONCLUSION

An analysis of the adverse effects and eventual failure of oral L-DOPA therapy—the current treatment of Parkinson's disease—leads to the conclusion that techniques which provide reinnervation of extensive, or functionally selective areas of striatum will fare better than simple drug infusion systems. In animal models, adrenal medulla and SN grafts have very similar behavioral effects. Nevertheless, embryonic SN at least has potential advantages over other dopaminergic donor tissues, and multiple-site transplantation of tissue or cell suspensions may provide better selective or regional reinnervation than single-site transplants.

REFERENCES

1. McDowell FH, Cedarbaum JM (1987). The extrapyramidal system and disorders of movement. In Baker AB, Joynt RJ (eds): "Clinical Neurology," Philadelphia: Harper and Row, ch 38, p 19.
2. Hornykiewicz O (1966). Metabolism of brain dopamine in human parkinsonism: Neurochemical and clinical aspects. In Costa E, Cote LJ, Yahr MD (eds): "Biochemistry and Pharmacology of the Basal Ganglia," New York: Hewlett, p 171.
3. Sweet RD, McDowell FH (1975). Five years treatment of Parkinson's disease with levodopa: Therapeutic results and survival of 100 patients. Ann Intern Med 83:456.
4. Kurlan R (1988). International symposium on early dopamine agonist therapy of Parkinson's disease. Arch Neurol 45:204.
5. Hood TW, Domino MD, Greenberg HS (1986). Reversal of parkinsonism in the MPTP rhesus monkey model by intraventricular L-dopa methyl ester. (abstr) Congress of Neurological Surgeons, 36th annual meeting, New Orleans.
6. Logan SR, Hargraves RW, Freed WJ (1988). Continuous intraventricular versus intrastriatal administration of L-dopa in rats with unilateral substantia nigra lesions. (abstr) American Association of Neurological Surgeons, 56th annual meeting, Toronto.
7. Hargraves RW, Freed WJ (1987). Chronic intrastriatal dopamine infusions in rats with unilateral lesions of the substantia nigra. Life Sci 40:959.

8. Markstein R (1981). Neurochemical effects of some ergot derivatives: A basis for their antiparkinson actions. J Neural Transm 51:39.
9. Barbeau A (1974). The clinical physiology of side effects in long-term L-dopa therapy. Adv Neurol 5:347.
10. Lesser RP, Fahn S, Snider SR (1979). Analysis of the clinical problems in parkinsonism and the complications of long-term levodopa therapy. Neurology 29:1253.
11. Leenders KL, Palmer AJ, Quinn N, Clark JC, Firnau G, Garnett ES, Nahmias C, Jones T, Marsden CD (1986). Brain dopamine metabolism in patients with parkinson's disease measured with positron emission tomography. J Neurol Neurosurg Psychiatry 49:853.
12. Freed WJ, Morihisa JM, Spoor E, Hoffer BJ, Olson L, Seiger R, Wyatt RJ (1981). Transplanted adrenal chromaffin cells in rat brain reduce lesion-induced rotational behavior. Nature 292:351.
13. Freed WJ (1983). Functional brain tissue transplantation: Reversal of lesion-induced rotation by intraventricular substantia nigra and adrenal medulla grafts, with a note on intracranial retinal grafts. Biol Psychiatry 18:1205.
14. Madrazo I, Drucker-Colin R, Diaz V, Martinez-Mata J, Torres C, Becerril JJ (1987). Open microsurgical autograft of adrenal medulla to the right caudate nucleus in two patients with intractable Parkinson's disease. N Engl J Med 316:831.
15. Becker JB, Freed WJ (1988). Neurochemical correlates of behavioral changes following intraventricular adrenal medulla grafts: intraventricular microdialysis in freely moving rats. Prog Brain Res (in press).
16. Freed WJ, Cannon-spoor EH, Krauthamer E (1986). Intrastriatal adrenal medulla grafts in rats: Long-term survival and behavioral effects. J Neurosurg 65:664.
17. Stromber I, Herrera-Marschitz M, Hultgren L, Ungerstedt U, Olson L (1984). Adrenal medullary implants in the dopamine-denervated rat striatum. I. Acute catecholamine levels in grafts and host caudate as determined by HPLC-electrochemistry and fluorescence histochemical image analysis. Brain Res 297:41.
18. Patel-Vaidya U, Wells MR, Freed WJ (1985). Survival of dissociated adrenal chromaffin cells of rat and monkey transplanted into rat brain. Cell Tissue Res 240:281.
19. Backlund EO, Granberg PO, Hamberger B, Knutsson E, Martenson A, Sedvall G, Seiger A, Olson L (1985). Transplantation of adrenal medullary tissue to striatum in parkinsonism: First clinical trials. J Neurosurg 62:169.

20. Stromberg I, Herrera-Marschitz M, Ungerstedt U, Ebendal T, Olson L (1985). Chronic implants of chromaffin tissue into the dopamine-denervated striatum: Effects of NGF on graft survival, fiber growth, and rotational behavior. Exp Brain Res 60:335.

21. Freed WJ, Spoor HE, de Beaurepaire R, Greenberg J, Schwarz S (1987). Embryonic substantia nigra grafts: Factors controlling behavioral efficacy and reinnervation of the host striatum. Ann NY Acad Sci 495:581.

22. Nieto-Sampedro M, Lewis E, Cotman C, Manthorpe M, Skaper S, Barbin G, Longo F, Varon S (1982). Brain injury causes time-dependent increase in neuronotrophic activity at the lesion site. Science 217:860.

23. Freed WJ, Patel-Vaidya U, Geller HM (1986). Properties of PC12 pheochromocytoma cells transplanted to the adult rat brain. Exp Brain Res 63:557.

24. Dichter MA, Tischler AS, Greene LA (1977). Nerve growth factor-induced increase in electrical excitability and acetylcholine sensitivity of a rat pheochromocytoma cell line. Nature 268:501.

25. Prochiantz A, DiPorzio U, Berger A, Glowinski J (1979). In vitro maturation of mesencephalic dopaminergic neurons from mouse embryos is enhanced in the presence of their striatal target cells. Proc Natl Acad Sci USA 76:5387.

26. Hoffer B, Rose G, Gerhardt G, Stromberg I, Olson L (1985). Demonstration of monoamine release from transplant-reinnervated caudate nucleus by in vivo electrochemical detection. In Bjorklund A, Stenevi U (eds), "Neural Grafting in the Mammalian CNS," Amsterdam: Elsevier, p 437.

27. Perlow MJ, Freed WJ, Hoffer BJ, Seiger A, Olson L, Wyatt RJ (1979). Brain grafts reduce motor abnormalities produced by destruction of the nigrostriatal dopamine system. Science 204:643.

28. Schwarz SS, Freed WJ (1987). Brain tissue transplantation in neonatal rats prevents a lesion-induced syndrome of adipsia, aphagia and akinesia. Exp Brain Res 65:449.

29. Wuerthele SM, Freed WJ, Olson L, Morihisa J, Spoor L, Wyatt RJ, Hoffer BJ (1981). Effect of dopamine agonists and antagonists on the electrical activity of substantia nigra neurons transplanted into the lateral ventricle of the rat. Exp Brain Res 44:1.

30. Freed WJ, Ko GN, Niehoff DL, Kuhar MJ, Hoffer BJ, Olson L, Cannon-Spoor HE, Morihisa JM, Wyatt RJ (1983). Normalization of spiroperidol binding in the denervated rat striatum by homologous grafts of substantia nigra. Science 222:937.

31. Freund TF, Bolam JP, Bjorklund A, Stenevi U, Dunnett SB, Powell JF, Smith AD (1985). Efferent synaptic connections of grafted dopaminergic neurons reinnervating the host neostriatum: A tyrosine hydroxylase immunocytochemistry. J Neurosci 5:603.
32. Fahn S, Libsch LR, Cutler RW (1971). Monoamines in the human neostriatum. J Neurol Sci 14:427.
33. Agid Y, Javoy-Agid F (1985). Peptides and Parkinson's disease. Trends Neurosci 8:30-35.
34. Boller F, Mizutani T, Roessmann U, Giambetti P (1980). Parkinson disease, dementia and Alzheimer disease: Clinicopathological correlations. Annal Neurol 7:329.
35. Dunnett SB, Bjorklund A, Stenevi U Iverson SD (1981). Grafts of embryonic substantia nigra reinnervating the ventrolateral striatum ameliorate sensorimotor impairments and akinesia in rats with 6-OHDA lesions of the nigrostriatal pathway. Brain Res 229:209.
36. Dunnett SB, Bjorklund A, Schmidt RH, Stenevi U, Iverson SD (1983). Acta Physiol Scand (Suppl) 522:39.
37. Wyatt RJ, Freed WJ (1985). Central nervous system grafting. In Wilkins RH, Rengachary S (eds): "Neurosurgery," New York: McGraw Hill, p 2546.

The views expressed herein reflect the authors, and are not those of the Department of Defense.

Tissue Engineering, pages 283–288
© 1988 Alan R. Liss, Inc.

MORPHOLOGICAL CORRELATES OF CHROMAFFIN CELL TRANSPLANTS IN CNS PAIN MODULATORY REGIONS [1]

George D. Pappas and Jacqueline Sagen

Dept. of Anatomy and Cell Biology, University of Illinois at Chicago, Chicago, Illinois 60612

ABSTRACT   Electron microscopic studies were undertaken to correlate the significant changes in pain sensitivity with the neural interactions of the host and adrenal medullary tissue grafts (or cultured, isolated chromaffin cell implants). The graft becomes encapsulated with collagen, while the host CNS tissue develops layers of glial processes outlining the graft. Nevertheless no absolute host-graft barrier is formed. We can find both neuronal and glial processes throughout the graft.   The vascular bed of the grafts allow some substances such as horseradish peroxidase to circumvent the blood-brain barrier and enter the extracellular space of the surrounding CNS parenchyma. Neuronal processes originating from the host tissue form synaptic junctions with the implanted chromaffin cells in the periaqueductal gray (PAG), but not when they are implanted in the subarachnoid space of the spinal cord.   The functional significance of the newly formed synapses in the PAG is unclear, since potent analgesia can be induced in animals with spinal cord subarachnoid grafts where we have not found synapses.

Adrenal medullary chromaffin cells contain and release several neuroactive substances, including catecholamines and opioid peptides.

[1] Supported by NIH Grants GM37326 & NS25054

The transplantation of chromaffin cells into CNS pain modulatory regions sensitive to these agents provides a locally available source of neuroactive substances for the relief of pain. We have shown that pain sensitivity can be reduced by transplanting chromaffin cells to the parenchyma of the rat midbrain periaqueductal gray (PAG) or in the dorsal subarachnoid space in the lumber enlargement of the rat spinal cord (1,2,3).

Our findings indicate that these transplants can function in the CNS environment and have the ability to integrate with the host CNS to some extent (3,4). Astrocytic cell processes delineate the host parenchyma. Facing this glial "shield" is a thick collagenous layer identifying the adrenal medullary graft. Although there is a clean delineation between the host and the graft, this does not appear to act as an absolute barrier to graft-host interaction. Glial and neuronal processes are readily found in the grafted tissue (4).

Neuritic processes are found to surround the surface of the chromaffin cells (Figs. 1,2,). In fact, neuronal processes originating from the host CNS form synaptic contacts with chromaffin cells not only in the adrenal medullary grafts but also when isolated cells from primary cultures are implanted. The behavioral relevance of these synaptic contacts is unclear, since similar implants into the spinal cord subarachnoid space, which also induce potent analgesia, do not contain synapses.

Non-syngeneic grafts in the brain and spinal cord survive because the CNS is considered to be "immunologically privileged" (5). Another important survival factor of the graft depends on its becoming vascularized within 8 hours following transplantation (6). We have transplanted into rats pieces of rat adrenal medullary tissue, isolated bovine chromaffin cells from primary cultures and a pheochromocytoma cell line culture (PC12) derived from a single clonal line of the rat (7). The endothelial cell lining of the capillaries in all the implants is of the

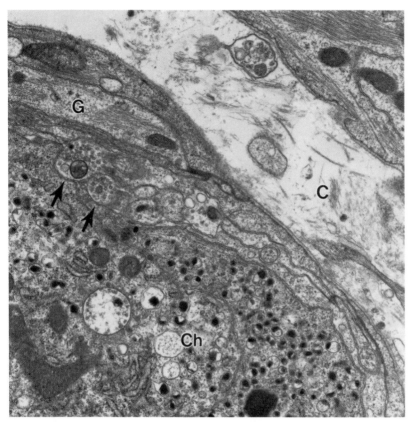

FIGURE 1.   Electron micrograph of a section of a portion of a chromaffin cell (Ch) in a graft surrounded by glial and neuronal processes of host origin.   Note, at arrows, two neuronal processes containing small clear presynaptic vesicles.   X16,000.

attenuated and fenestrated type   (Fig. 3).   This type of vascular bed differs from that found in the surrounding host CNS parenchyma (8).   Both the periaqueductal gray and spinal cord have a continuous non-fenestrated endothelium, which characterizes   morphologically   the   permeability properties   of   the   CNS   (i.e.   the   blood-brain barrier).   The   vascular   bed   of   the   grafts   may

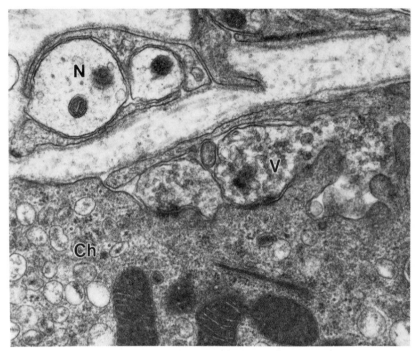

FIGURE 2.   EM of a section showing a portion
of a chromaffin cell with two synaptic junctions
on its surface.   The presynaptic processes
contain clusters of mostly small clear vesicles
(V).   Other neuronal processes (N) surrounded by
glial processes can be seen.   This section was
taken from a rat adrenal medullary graft implanted 8 weeks earlier in the periaqueductal gray.
X20,000. (Taken from Pappas and Sagen, 8)

circumvent   the   blood-brain   barrier   of   the
surrounding host CNS.   The   intravascular injection of the protein marker horseradish peroxidase
(HRP) first enters the extracellular space of the
graft parenchyma and then the extracellular space
of the host CNS (9).   We can conclude that the
permeability properties of the vascular endothelial cells are determined by the cells and tissue
of their immediate local environment and not by
their CNS   origin.   Implanted   chromaffin   cells,
like   other   neuroendocrine   cells,   recapitulate

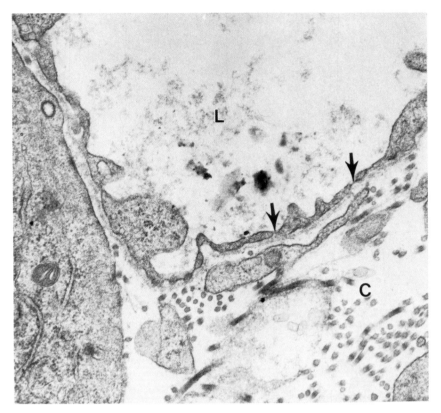

FIGURE 3. EM of a portion of an attenuated capillary endothelial cell within a cluster of PC-12 cells in the rat spinal cord parenchyma. Note the pores or fenestrae (at arrows). Collagen fibers (C) are found in an enlarged adlumenal area. L = Lumen of capillary. X30,000.

their original *in situ* environment, by having a fenestrated capillary endothelium.

ACKNOWLEDGEMENT

The Electron Microscope Facility of the Research Resources Center, University of Illinois at Chicago was used in conducting this study.

REFERENCES

1. Sagen, J., Pappas, G.D., and Perlow, M.J., (1986). Adrenal medullary tissue transplants in the rat spinal cord reduce pain sensitivity. Brain Res 384:189.

2. Sagen, J., Pappas, G.D., and Pollard, H.B., (1986). Analgesia induced by isolated bovine chromaffin cells implanted in rat spinal cord. Proc Natl Acad Sci 83:7522.

3. Sagen, J., Pappas, G.D., and Perlow, M.J., (1987). Alterations in nociception following adrenal medullary transplants into the rat periaqueductal gray. Exp Brain Res 67:373.

4. Sagen, J., Pappas, G.D., and Perlow, M.J., (1987). Fine structure of adrenal medullary grafts in the pain modulatory regions of the rat periaqueductal gray. Exp Brain Res 67:380.

5. Medawar, P.B., (1948). Immunity to homologous grafted skin. II. The fate of skin homografts transplanted to the brain, to subcutaneous tissue, and to the anterior chamber of the eye. Brit J Exp Pathol 29:58.

6. Krum, J.M. and Rosenstein, J.M., (1987). Patterns of angiogenesis in neural transplant models: I. Autonomic tissue transplants. J Comp Neurol 258:420.

7. Pappas, G.D. and Sagen, J., (1986). Fine structure of PC12 cell implants in the rat spinal cord. Neurosci Lett 70:59.

8. Pappas, G.D. and Sagen, J., (1986). The fine structure and permeability of endothelial cells in vascularized tissue and cell implants in the CNS. Anat Rec 214:96.

9. Pappas, G.D., and Sagen, J., (1988). The fine strucutre of chromaffin cell implants in the pain modulatory regions of the rat periaqueductal gray and spinal cord. In: "Transplantation in the Mammalian CNS." Sladek, R. and Gash, D. (eds) Elsevier Press, Amsterdam, in press.

Tissue Engineering, pages 289–294
© 1988 Alan R. Liss, Inc.

PAIN REDUCTION BY CHROMAFFIN CELL GRAFTS IN THE CNS [1]

Jacqueline Sagen and George D. Pappas

Dept. of Anatomy and Cell Biology, University of Illinois at Chicago, Chicago, IL 60612

ABSTRACT   Transplants of adrenal medullary tissues or isolated chromaffin cell suspensions into regions of the neuraxis involved in pain modulation can induce analgesia.   This analgesia most likely results from the stimulated release of neuroactive substances such as opioid peptides and catecholamines from the grafted cells.   Results of this work have suggested a new possible approach towards the management of intractable pain.

INTRODUCTION

In recent years, we have witnessed an explosion in the field of neural transplantation due to the discovery that transplants into the brain have the potential to alter behavioral function in the host (see 1 for review). Our laboratory has been interested in the possibility of using neural grafts to alleviate pain. Towards this goal, we have been transplanting adrenal medullary tissue into specific pain modulatory regions of the neuraxis, including the dorsal spinal cord subarachnoid space and the midbrain periaqueductal gray. The choice of adrenal medullary tissue for transplantation is based on several properties of chromaffin cells contained in the adrenal medulla. Most importantly, these cells synthesize high

[1] Supported by NIH Grants GM37326 & NS25054

neuroactive substances which can induce potent analgesia when injected locally into CNS pain modulatory regions (2,3). In addition, the release of these substances can be augmented by stimulation of cell surface nicotinic receptors using nicotinic agonists. To date, donor chromaffin cells have been obtained from rat, bovine, and human adrenal medulla with similar results (4-7). The possibility of using xenografts extends the range of donor sources that may ultimately be used clinically.

METHODS AND RESULTS

Adrenal medullary tissue for transplantation was obtained from adult Sprague-Dawley derived rats. Following careful dissection, adrenal medullary tissue was transplanted either into the rat spinal cord subarachnoid space at the lumbar enlargement, or into the midbrain periaqueductal gray (PAG). Control rats received equal volumes of either sciatic nerve tissue, heat-killed adrenal medullary tissue, or gelfoam. Cell suspension implants included isolated and cultured bovine chromaffin cells and PC12 cells, a clonal pheochromocytoma cell line. Cells were also implanted in the spinal cord subarachnoid space or into the PAG. Details of these surgical procedures have been published previously (4-7).

Pain sensitivity was assessed using three standard analgesiometric tests sequentially: the tail flick test, paw pinch test, and hot plate test. The use of these three tests allows the measurement of responses to both thermal and mechanical stimuli, as well as both reflexive and integrative pain responses. Details of analgesiometric testing procedures have been described elsewhere (4,5).

Changes in pain sensitivity following the transplantation of adrenal medullary tissue into the spinal cord subarachnoid space are illustrated in Fig. 1. This data was obtained from 8 week old grafts. The injection of low doses of nicotine (0.1 mg/kg, s.c.) produced potent analgesia in animals with adrenal medullary implants. The

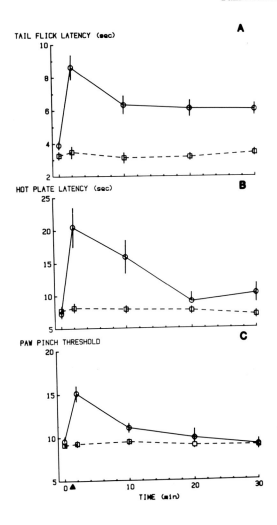

FIGURE 1. Effect of spinal cord adrenal medullary transplants on pain sensitivity. The ordinate is the threshold for response to noxious stimuli as assessed by the three analgesiometric tests. The arrowhead indicates the point at which nicotine was injected. Open circles: animals with adrenal medullary transplants in the spinal cord (N=12); Open squares: animals with control transplants in the spinal cord. Taken from Sagen et al. (4).

onset of this analgesia is rapid (within 2 minutes) and lasts approximately 20 min. before returning to baseline pain levels. This same dose of nicotine has no effect on pain sensitivity in animals with control implants.

The contribution of opioid peptides to the induction of analgesia by these transplants was assessed using the opiate antagonist naloxone. Following the induction of analgesia with nicotine, animals with spinal cord adrenal medullary transplants were injected either with naloxone (2 mg/kg, s.c.) or saline vehicle. Naloxone reversed the analgesia to baseline levels, while saline had no effect.

The results of these studies indicated that it is possible to reduce pain sensitivity with transplants of adrenal medullary tissue in the spinal cord subarachnoid space. This reduction in pain sensitivity is most likely due to the stimulated release of opioid peptides from grafted cells since it can be reversed by naloxone. In addition, we have found that catecholamines also play a role in the induction of this analgesia, since it is attenuated by adrenergic antagonist phentolamine. Recent work in this laboratory has involved the more direct measurement of the release of neuroactive substances from the adrenal medullary transplants using spinal cord superfusion techniques. CSF samples were collected from animals with adrenal medullary or control transplants, and analyzed for met-enkephalin release using radioimmunoassay. Results are shown in Table 1.

These results show that in animals with adrenal medullary spinal cord transplants, release of met-enkephalin is nearly doubled. In addition, this release is further augmented by nicotine stimulation in transplanted, but not control animals.

TABLE 1
MET-ENK RELEASE FROM ADRENAL MEDULLARY GRAFTS

Transplant Group	Basal Met-Enkephalin	Nicotine-stimulated Enkephalin
Adrenal Medullary Implants	14.8 +\- 1.8	23.2 +\- 2.6
Control Implants	7.8 +\- 1.2	6.9 +\- 1.3

Met-enkephalin levels are expressed in pg/ml +\- S.E.M. for 5 minute samples

DISCUSSION

The results of these studies have indicated that it is feasible to transplant opioid peptide producing cells into the CNS to alter pain sensitivity. In addition to the spinal cord subarachnoid space, similar adrenal medullary transplants produce analgesia when placed in the midbrain PAG (7). We have also found that the transplantation of isolated suspensions of bovine chromaffin cells at either of these sites effectively reduce pain sensitivity (5,7). The advantage of using cell suspensions rather than solid tissue is that the graft is much more homo- geneous; i.e. the chromaffin cell suspensions are devoid of other cell types present in solid adrenal medullary tissue, including endothelial cells, fibroblasts, and ganglion cells. An additional advantage of bovine chromaffin cells is that they produce much greater quantities of opioid peptides. The survival of these bovine cells in the rat CNS supports the notion of CNS "immunologic privilege" (8), although this has still not been well characterized. Should it be possible to use cultured cells from genetically disparate sources as tissue for CNS grafts in clinical situations, the potential availabilty of such grafts will expand enormously. In addition, it may be possible to genetically engineer cells for transplantation. In this regard, cell lines which have been transfected with proenkephalin genes and produce high levels of opioid peptides may be ideal for transplantation in the management of intractable pain.

REFERENCES

1.   Azmitia EC and Bjorklund A (1987). "Cell and Tissue Transplantation into the Adult Brain." New York: New York Academy of Sciences, 813 p.
2.   Yaksh TL and Reddy SVR (1981). Studies in the primate on the analgesic effects associated with intrathecal actions of opiates, alpha-adrenergic agonists and baclofen. Anesthesiology 54:451.
3.   Yaksh TL, Yeung JC, and Rudy TA (1976). Systematic examination in the rat of brain sites sensitive to the direct application of morphine: Observation of differential effects within the periaqueductal gray. Brain Res. 114:83.
4.   Sagen J, Pappas GD, and Perlow MJ (1986). Adrenal medullary tissue transplants in the rat spinal cord reduces pain sensitivity. Brain Res. 384:189.
5.   Sagen J, Pappas GD, and Pollard HB (1986). Analgesia induced by isolated bovine chromaffin cells implanted in rat spinal cord. Proc. Natl. Acad. Sci. USA 83:7522.
6.   Sagen J, Pappas GD, and Perlow MJ (1987). Fine structure of adrenal medulary grafts in the pain modulatory regions of the rat periaqueductal gray. Exp. Brain Res. 67:380.
7.   Sagen J, Pappas GD, and Perlow MJ (1987). Alterations in nociception following adrenal medullary transplants into the rat periaqueductal gray. Exp. Brain Res. 67:373.
8.   Freed WJ (1983). Functional brain tissue transplantation: Reversal of lesion-induced rotation by intraventricular substantia nigra and adrenal medulla grafts, with a note on intracranial retinal grafts. Biol. Psychiat. 18:1205.

Tissue Engineering, pages 295–296
© 1988 Alan R. Liss, Inc.

Summary/Discussion

Functions of Cells and Tissue Implants

George Pappas

Important neurological disorders were discussed in terms of tissue engineering approaches to Parkinson's and Alzheimer's disease and to intractable pain. Approaches to transplanting embryonic and adult tissue into the brain to replace non-functioning tissues was discussed. This implantation approach was suggested for Alzheimer's and Parkinson's diseases for the alleviation of these symptoms by substituting "healthy" tissue for the non-functioning nerve cells. Implantation in the areas of the central nervous system, where the brain perceives pain, is also a promising approach to the alleviation of severe chronic and phantom pain.

Dr. Freed reviewed findings from his and other laboratories on the use of catecholaminergic brain grafts in animals with lesions in the substratia nigra (SN). Two kinds of neural implants -embryonic nigral tissue and adult adrenal medulla are placed into the striatum - as a replacement for the missing nerve endings of the SN. Both of these implants apparently "correct" motor abnormalities brought about by the experimentally produced deficit.

Transplantation of chromaffin cells into the central nervous system of rats provided a good example of how cells and tissues may be used in humans predictably in the next ten years. Before this can happen, a broad range of problems needs to be solved as is the case with other transplantable cell systems. One practical and useful spin-off of this research will be a better understanding of functional cellular elements defining the blood brain barrier.

Drs. Pappas and Sagen reported on their work on pain reduction by chromaffin cell grafts in the CNS. Transplants of adrenal medulary tissues or isolated chromaffin cell suspensions into regions of the neuraxis involved in pain modulation can induce analgesia. This analgesia most likely results from the stimulated release of peptides and catecholamines from the grafted cells. Results of their work have suggested a new possible approach towards the management of intractable pain.Electron microscopic studies show that the graft becomes encapsulated with collagen, while the host CNS tissue develops layers of glial processes outlining the graft. However, these layers do not prevent neuronal and glial processes from entering the grafted area.

Tissue Engineering, pages 297–298
© 1988 Alan R. Liss, Inc.

VII. The Hematopoietic System

Introduction. . . . . .Randall Swartz

The hematopoietic system is responsible for diverse functions in the immune and circulatory systems, including oxygen transport, artery repair and control of diseases of foreign origin (virus and bacteria and toxic substances) as well as such regulatory aberrations as cancer and autoimmune disfunction.

All the elements (cell types) in the system are derived from the pleuripotent stem cells residing on the bone marrow. All are non-anchorage dependent and all share complex multistop differentiation to their final, highly specialized and tightly, interactively regulated functions. Many disease states derived directly from upsets in these complex interactions.

Two fields of development in the mid 1970's - monoclonal antibodies and recombinant DNA technology - led directly to rapid advances in our understanding of these systems. Monoclonal antibodies against unique markers in the cellular immune system are of particular utility in understanding its action and regulation. Recombinant DNA technology has made available dozens of biological response modifiers, including thelymphokines and cytokines which are hormones responsible for regulating the hematopoietic system.

In some cases, it is necessary to remove cells from the patient and manipulate them ex-vivo for reinfusion. One example would be bone marrow transplantation in chemotherapy which has proved utility in the control of metastatic disease. Promising results have been achieved in the control of previously intractable forms of cancer using so called "LAK" and "TIL" cells, but these therapeutic modalities are cumbersome, difficult to control and expensive.

Engineering approaches are needed in several areas: cell sampling, separation and identification of active cell types from among the mixtures of patient derived cells, cost effective and clinically acceptable activation and expansion of these to clinically useful quantities, while controlling the subpopulations so that the correct mixture is obtained, understand the modulation of subpopulations by lymphokines and cytokines, development of bioreactors to support cell expansion sufficient for therapy, and enable quantitation of the interaction of cell subsets with each other (these cells produce many of their own regulators and growth factors). Systems for the storage and administration of cells would also benefit from engineering support.

Sufficient information is now available to support development of a mechanistic model of the regulation of the human cellular immune system response to cancer and other diseases. Since these efforts would significantly aid the development of therapeutic modalities, they should be investigated.

Oncologists, immunologists, hemotologists, and specialists inautoimmune disease working on these areas are making progress, but they are clearly hampered by a lack of engineering approach in both understanding of their results quantitatively as well as extending them to routine, cost effective therapies.

Since the initial therapies utilizing in vitro expansion of hematopietric cells, especially those of the immune system, are already in clinical trials and there is broad agreement as to the need for engineering solutions to improve delivery technology, this appears to represent a clear opportunity for multidisciplinary collaborations between immunologists and biochemical engineers. Biochemical engineers can also contribute significantly to developing convenient, cost effective bioreactors for cell expansion and in developing practical separations of individual immune cell types in order that their role in therapy can be determined and this understanding applied to improve therapy. They can also aid the immunologist in understanding and quantifying the complex regulatory interactions of the cellular immune system using quantitative, mechanistic remodeling techniques and in the design of more effective, less invasive therapeutic modalities combining cellular immunotherapy with lymphokine and cytokine therapy.

Tissue Engineering, pages 299–312
© 1988 Alan R. Liss, Inc.

ACTIVATION AND EXPANSION OF CELLS FOR ADOPTIVE IMMUNOTHERAPY[1]

Randall W. Swartz, Allan Haberman, Don DiMasi, Bruce Jacobson, Ana Lages, Mark Grise', Dimitri Nicolakis and George Truskey

Biotechnology Engineering Center, Tufts University, Medford, MA 02155

ABSTRACT   "Adoptive" cellular immunotherapy for cancer involves the reinfusion of a patient's own lymphocytes in order to induce an effective immune response. The cells are isolated, activated and expanded in several ways to prepare them for use. These experimental therapies have shown variable degrees of clinical response. However, the mechanisms of tumor regression are not understood.   The cells returned to the patient are populations of lymphocyte subsets and information on their interaction and relative significance is needed to design more effective therapies and to learn how to activate and expand the cells to the numbers required for therapy with control over the population. Lymphocytes have unique requirements for activation and expansion relating to cell density, O_2 tension, shear and other environmental factors and these will be discussed from the viewpoint of controlling growth, population mix and activation. Bioreactor and nutritional control options for increasing cell numbers to therapeutically useful quantities and for controlling the physiological state of the cells will be discussed. Bioreactor and nutritional control approaches may be relevant to other areas of "tissue engineering".

INTRODUCTION

There are many metastatic tumors for which no effective therapy exists. There is no histological class of tumor that is resistant to LAK mediated killing in vitro. These two facts form the basis for the great interest in adoptive therapy. Adoptive cellular immunotherapy (ACI) for cancer involves administration of cells with anti-tumor reactivity to patients, in order to induce tumor regression. S. A.

[1] This work is supported by a grant from the National Science Foundation Directorate for Engineering. Advice from Drs. James Kurnick and James Mier is gratefully acknowledged. Contribution of rIL-2 by Amgen has been essential to this project.

Rosenberg *et al.* (New England J. Med., 313:1485, 1986) have developed a treatment method which involves obtaining peripheral blood mononuclear cells (PBMC) from cancer patients and incubating them in vitro in the presence of interleukin-2 (IL-2). This results in the generation of a class of lymphocytes known as lymphokine activated killer (LAK) cells.

The "LAK phenomenon" is defined as the generation in IL-2-treated PBMC of cells having non-MHC (major histocompatibility complex) restricted cytotoxic activity against a wide variety of fresh tumor cells and against both NK (natural killer) resistant and NK sensitive tumor cell lines. The cells responsible for the in vitro cytotoxicity are primarily derived from NK cells, but some are derived from T-cell progenitors (R.B. Herberman *et al.*, Immunology Today,178:1987). IL-2 is required not only for generation of LAKs from PBMC, but also for the growth and viability of LAKs. PBMCs cultured in IL-2, commonly referred to as "LAK cells", are complex populations, which include a wide variety of lymphocytes in addition to the non-specific cytotoxic lymphocytes which are defined as having in vitro "LAK" activity.

Rosenberg and coworkers (A. Mazumder and S.A. Rosenberg, J. Exp. Med.,159: 495, 1984; J.J. Mule' *et al.*, Science, 225: 1487, 1984; J. J. Mule' *et al.*, J. Immunol., 135: 646, 1985; R. Lafreniere *et al.*, Cancer Res., 45: 3735, 1985) have demonstrated that adoptive transfer of mouse LAK cells, administered together with IL-2, can mediate regression of metastatic tumors in mice.

Rosenberg's group and others (S.A. Rosenberg,*et al.*, New England J. Med., 313:1485, 1986; S.A. Rosenberg, Immunology Today, 9:58, 1988; S. A. Rosenberg, *et al.*, New England J. Med, 15: 889,1987) then applied this experimental therapy to humans, administering autologous LAKs, together with IL-2, to cancer patients. Most success (partial to complete responses) has been achieved with metastatic renal cell carcinoma. Significant numbers of partial to complete responses have also been seen with melanomas. However, there is some evidence to suggest that IL-2 alone (without LAK cells) may be efficacious with melanomas. Rosenberg and coworkers have also reported some responses with colorectal cancer and non-Hodgkin's lymphoma, but very few patients have been examined.

Other approaches to ACI, involving cytotoxic T-cells which are more specific for a given tumor as opposed to the broadly tumor-specific LAKs, are also under development. Several laboratories, including Rosenberg's laboratory (S.A. Rosenberg, *et al.* Science, 233: 1318, 1986; L.M. Muul, *et al.*, J. Immunol., 138:989, 1987; S.L. Topalian, *et al.*, J. Immunol. Methods, 102: 127, 1987) and the laboratory of James T. Kurnick at Massachusetts General Hospital (J.T. Kurnick, *et al.* Clin. Immunol. Immunopathol., 38: 367,1986) are developing an alternative method of ACI based on use of tumor-infiltrating lymphocytes (TILs). TILs are defined operationally as lymphocytes which infiltrate tumors and grow when tumor explants are cultured in media containing IL-2. TILs are complex populations, consisting largely of T-cells, some of which have been shown to exhibit in vitro cytotoxicity for autologous tumor cells. Cytotoxicity exhibited by TIL cells may be of two types: non-MHC restricted, non-specific cytotoxicity similar to that shown by LAK cells, and MHC-restricited cytotoxicity which in some cases has been shown to be specific for autologous tumor. In some cases when autologous tumor and TILs were available for testing, the tumor-derived lymphocytes were shown to be about 50-100-fold more effective in lysing autologous tumor cells in vitro than were autologous IL-2 responsive blood lymphocytes (produced by a method similar to Rosenberg's LAKs). In Rosenberg's mouse model (S.A. Rosenberg, Science, 233: 1318, 1986), this increased in vitro cytotoxicity correlated with an increased effectiveness in regression of in vivo established metastases. This, together with a lower level of clinical performance by the LAK therapy than expected as a result of preliminary clinical trials, has led to a strong interest in the TIL therapy by several clinical immunology groups.

The ability of clinical groups to administer these therapies to large numbers of patients is severely limited by cost and logistical factors. This is due in large part to the inability to readily expand the activated human lymphocytes in vitro in numbers large enough for effective therapy. For example, Rosenberg's LAK therapy requires 4×10^{10} to 2×10^{11} cells per patient, given in multiple treatments. In order to approach these numbers, Rosenberg's group and the other laboratories conducting clinical trials of the therapy must obtain large numbers of PBMC directly from patients. This is done by 1) pre-treatment of patients with large doses of IL-2 over a 5-day period during the first week of the treatment cycle (which has been found to increase total numbers of peripheral blood lymphocytes and LAK precursors), followed by 2) extensive leukapheresis over five successive days during week two of the treatment cycle. They culture the harvested cells with IL-2 in roller bottles, using large volumes (110-150 liters/patient) of medium containing human serum. During week three of the treatment cycle, they infuse the cells, together with IL-2, into patients over 5 successive days. This treatment involves considerable side effects due to IL-2 administration. The ability to readily culture and expand LAKs would result in either elimination or lessening of the need for leukapheresis, thus lowering costs and improving the logistics of treatment so that more patients could be treated. The ability to expand LAKs could also eliminate the need for pre-treatment of patients with IL-2, thus eliminating the side effects which may occur during the pre-treatment period. In the case of TIL cells, one must either obtain large enough tumor samples such that one can grow enough cells for therapy without reactivation (the approach reported by Rosenberg's group), or one must reactivate cells using 5-10 times the number of "feeder cells" (irradiated PBMC obtained from fresh blood) plus the lectin phytohemagglutinin (PHA) (an approach used by Dr. James Kurnick of MGH). One cannot always obtain such a large tumor sample, and reactivation of TILs with feeder cells and PHA, and culture in T-flasks, etc. is cumbersome, expensive, and does not allow for the treatment of large numbers of patients.

CURRENT METHODS OF CELL PRODUCTION

Lymphokine Activated Killer (LAK) Cells

LAKs are made by suspending PBLs in RPMI 1640 plus 2 to 10% heterologous donor human serum plus 100 to 1000 units recombinant IL-2 (L.M. Muul *et al.*, J. Immunol. Methods, 88: 265, 1986.] LAK cultures used in clinical trials are 4-5 day cultures. There is no net expansion of lymphocytes. Such cultures consist of non-activated T-cells plus a variety of lymphocyte subsets which have become activated when cultured in IL-2. A minority of these subsets are thought to be responsible for the major "LAK" cytotoxic activity (R.B. Herberman, *et al.*, Immunology Today, 8: 178, 1987). However, it is not known which of the subsets are responsible for tumor regression, either in humans or in mouse models.

Tumor Infiltrating Lymphocytes (TIL)

To isolate TILs, a tumor biopsy is removed, minced and sometimes treated with tissue degrading enzymes such as collagenase. The biopsy is placed in a well (6-well plate) and cultured in RPMI1640 plus 5 to 10% human serum (single donor) plus 100-1000 units/ml of IL-2. The lymphocytes expand and destroy the remaining tumor cells. Once the tumor is destroyed or the T-cells are transfered to fresh medium (via Ficoll or aspiration), they will proliferate for a limited time (circa 7 to 10 doublings). If required, they are then reactivated using feeder cells plus PHA. Five to ten times as many feeders (irradiated PBLs) are required as the TIL cells to be reactivated. Once reactivated, expansion may continue for another circa 7 to 10 doublings.

The resulting culture contains a mixed population of circa 90% T-cells which is quite variable with respect to T-cell markers. The population may shift markedly during the expansion. Shifts in proportions of T-cells with surface markers CD4 (predominantly helper T-cells) and CD8 (predominantly cytotoxic and/or suppressor T-cells) during culture of T-cells in Rosenberg's laboratory (S.L. Topalian, *et al.*, J. Immunol. Methods, 102:127, 1987; L.M. Muul, *et al.*, J. Immunol., 138: 989, 1987) are illustrated in figure 2. Early passage TILs may also contain significant proportions of NK cells, at least some of which may have LAK activity. However, in many cases these NK cells are lost during culture. In addition to lymphocyte subsets, as defined by phenotypic markers, specific and non-specific in vitro cytotoxicity exhibited by TIL populations may also change during culture (See figure 1.).

Rosenberg's original mouse model studies (S.A. Rosenberg,*et al.*,Science, 233: 1318, 1986) indicated a correlation between the 50-100 fold greater in vitro cytotxicity of the TIL population as compared to LAKs with a corresponding greater in vivo effectiveness in regression of established metastases. However, as is the case with LAKs, no correlation has been confirmed for TILs between the population mix or in vitro cytotoxicity and clinical efficacy. Moreover, other evidence indicates that direct killing by infused cytotoxic cells is unlikely. Indium labeled LAK cells (S.A. Rosenberg *et al.*, <u>New England J. Med.</u>, <u>313</u>: 1485, 1985; J. Mier,pers.comm.) and TIL cells (J.T. Kurnick, pers. comm.) migrate primarily to liver, spleen, and lung capillary beds. There is no evidence of migration to individual metastases, even in cases where positive clinical responses are obtained. Many clinicians working with these systems favor an explanation for the LAK and TIL induced tumor regression in vivo which relies on indirect immune system regulation rather than direct lysis of tumor cells by infused LAKs or TILs. Because of the potential importance of suppressor T-cells in supression of either direct cytotoxicity or activation of other effector cells by infused TIL cells, as well as direct evidence in mouse models (S.A. Rosenberg, *et al.*, <u>Science.</u> <u>233</u>: 1318, 1986), some groups are evaluating the use of cymetidine (Tagamet) or cyclophosphamide to reduce suppressor population activity.

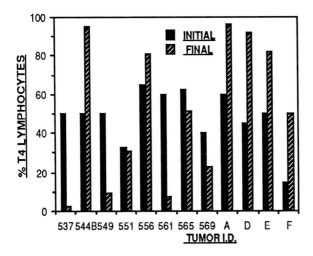

Figure 1. Population shifts observed during expansion of human TILs for therapy.   Data from twelve patients shows marked, unpredictable changes in subsets. Data derived from S.L.Topalian, *et al.*, <u>J. Immunol. Methods</u>, <u>102</u>: 127, 1987; and L.M. Muul, *et al.*, <u>J. Immunol.</u>, <u>138</u>:989, 1987.

LAK CELL EXPANSION

In our laboratory, we are interested in the study of long-term cultivation of active LAK populations as an approach leading to the production of the amounts of cells needed for therapy without extensive leukapheresis of patients. This general approach is also being investigated in other laboratories (A. A. Rayner *et al.*, Cancer, 55: 1327,1985; A.C Ochoa *et al.*, J. Immunol., 138: 2728, 1987; N. L. Vujanovic *et al.*, J. Exp. Med., 167: 15, 1988.) We take 50 cc of blood which yields approximately 10^8 PBMC. Cells are cultivated in 1000 u/ml IL-2, 10% human serum, RPMI1640, and penicillin/streptomycin for 4 days, yielding a "4-day LAK culture". These cells are then separated in discontinuous Percoll gradients (using 30%, 46%, and 55% Percoll layers). This yields 1-2 x 10^7 active Percoll separated low (sedimenting between the 30% and the 46% layers) and intermediate (between the 46% and 55% layers) density lymphocytes. The gradients remove dead cells and non-activated T-cells (which presumably are inactive in the standard LAK treatment.) Low and intermediate density lymphocytes are enriched in NK cells, but also contain considerable numbers of CD3-positive T-cells. Cells from these fractions are capable of growth in IL-2 containing media, and these populations are enriched in cells having large granular lymphocyte (LGL) morphology, and are highly clumped. Cells in the pellet are small, compact lymphocytes, which are >95% CD3-positive. The latter cells show little or no growth in IL-2 containing media.

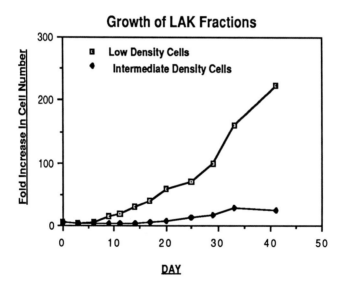

Figure 2. Long term growth of active lymphocytes derived from Percoll fractionation of a 4 day LAK culture.

The low and intermediate density cells are cultivated in 1000 u/ml IL-2, 10% human serum, RPMI1640, and pen/strep. We observe long term growth of this fraction as seen in figure 2. The cytotoxicty of one preparation of long-term cultured LAKs was equivalent to standard 4-5 day LAKs used in immunotherapy. Long-term cultured LAKs were found to consist mostly of CD3-positive T-cells, but to contain significant numbers of NK cells. We have obtained levels of expansion ranging from about 10-200 fold. Further studies on reactivation and expansion of our long-term LAK cultures are in progress.

TIL CELL EXPANSION

Approximately 10^{10} to 10^{11} cells are required for LAK or TIL therapy. In our laboratory we find that TIL cells will maintain active growth for circa 7 doublings following reactivation. The degree of expansion and thus the number of reactivations required depends upon the number of lymphocytes obtained from the tumor sample. In some cases this can be as large as 10^8 cells and no activation is required. In other cases only small tumor samples (yielding circa 10^6 TILs) are available, and two reactivations may be required. Proliferation rate and maintenance of activation may be influenced by the nutrient environment, concentration of toxic byproducts (e.g. lactate and NH_3), oxygen availability, and by growth factors (provided by serum or in defined media, as well as autocrine growth factors). Direct cell-cell contact may also be important.

Current clinical trials are being conducted using cells prepared via brute force approaches. Circa 100-150 liters of tissue culture medium containing up to 10% human serum and from 10^7 to 1.5×10^8 units of rIL-2 are required to prepare the cells of a single patient. In some cases, with non-optimal growth procedures, expansion can require 4 to 6 weeks and use dozens of laboratory scale flasks, poorly suited to monitoring or control of cultivation conditions and requiring intensive, skilled manual labor.

With only a single unit of blood or less for feeder cells (irradiated peripheral blood mononuclear cells- lymphocytes and monocytes- obtained using Ficoll) one can get clinical quantities of TILs using bioreactors. Feeder cells are dying (irradiated) cells required for reactivation along with the lectin PHA and IL-2. One unit (450 ml.) whole blood contains circa1.5×10^6 PBMCs per ml. One requires about 5 times the number of feeder cells as TIL cells to be reactivated. Thus, the number of TILs which can be reactivated is 1.35×10^8. During the first week of reactivation, TILs will undergo 3 doublings giving 1.08×10^9 TILs which can grow with a 24 to 48 hour doubling time. In 12 days there should be 6.9×10^{10}. This is sufficient for currently proposed therapies but very difficult with T-flasks.

Advantages of Cultivation in Bioreactors

In flasks- the density limit is roughly 1×10^6 cells/ ml and one T-175 flask holds circa 600 ml of media. To reach 5×10^{10} cells 100 flasks are needed. In the new Dupont bags, one can get 1×10^9 cells in 500 ml of media with a media change approximately every 6 hours but growth rates are slower. In the right bag or other bioreactor one should be able to greatly simplify this procedure operationally.

The influence of culture conditions on growth rate and on the number of doublings which may be achieved between reactivations is unknown. In our laboratory we have been studying the effects of growth media and culture conditions on proliferation of LAK and TIL preparations. With TIL cells we find that growth rates vary significantly depending on nutrient and waste product concentrations. For example, we find that even varying glucose concentration in fed batch culture had a significant effect upon growth rate and extent as seen in figure 3. As glucose concentration is increased, growth decreased in rate and extent. It was also observed that the extent of aerobic vs. anaerobic metabolism varies with nutrient conditions. Glutamine, the other primary carbon and energy source, was shown to significantly affect TIL growth as shown in figure 4.

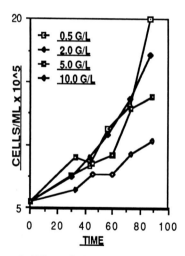

Figure 3. Effect of glucose on growth of TIL cells. Cells were cultivated in upright T flasks. At each sample time 40% to 60% of the media was replaced in order to minimize changes in the nutritional environment. The maximum rate and extent of growth was observed at a glucose concentration of 0.5 g/l, 1/4 the standard level.

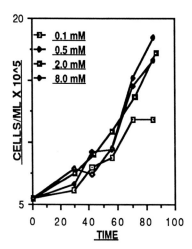

Figure 4. Effect of glutamine on TIL growth. Maximum growth was observed with glutamine maintained near 4x the standard starting level.

It has also been demonstrated that TIL cell cultures are influenced by physical factors. We were at first surprised to learn from our clinical colleagues that TIL cells grew well when grown in upright T flasks. We would have predicted that oxygen availability might have restricted cell proliferation under these conditions. It is now clear that this geometry, involving growth with the cells grouped together on the small bottom surface of the flask, enhances cell growth and maintenance of the activated state. Indeed, if the T-flask is laid on its side, the growth rate is significantly reduced. The experiment presented in table 1 quantifies this response for a 4.5 day cultivation of TIL.

Possible explainations for this phenomenon are that cell - cell contact is important or that the cells produce growth factors which are initially higher in concentration in the region of high density. We have observed that lymphocytes derived from TIL or LAK cultures grow poorly or not at all in traditional spinner type suspension cultures even when inoculated at relatively high densities.

TABLE 1
EFFECT OF CELL/SURFACE AREA ON GROWTH RATE[a]

media volume(ml)	initial value cells/cm^2	doublings
20	2.1 x 10^4	1.70
20	4.2 x 10^4	1.70
20	8.3 x 10^4	2.20
20	2.1 x 10^5	2.26
20	4.2 x 10^5	2.23
20	8.3 x 10^5	1.54
40	8.3 x 10^5	1.75
80	8.3 x 10^5	2.41

[a]Effect of cell #/surface area on growth rate. At low cell #/surface area, growth rate is negatively affected. The last three entries show that once this restriction is relieved, nutrient availability or dilution of toxic wastes, but not oxygen availability, restricts the rate of growth. Thus, above a threshold level of cells per surface area, other factors become important. It should be noted that total cell counts in these experiments are sufficiently low that oxygen limitation would not have been anticipated.

BIOREACTOR SELECTION

Bioreactor Designs

One goal of our research is to cultivate TIL cells at high cell density in order to simplify their expansion for therapy. In bioreactors of appropriate design we feel that growth rates may be maximized and the media volume required and media cost would be reduced, first because enhanced growth rate would shorten cultivation time and second because in the current procedure much of the nutrient content is wasted. At high cell density the cells would more efficiently utilize endogenously produced growth factors. If one or two bioreactors were used, perfusion of nutrient containing media would be feasible and if proved useful it would be possible to set up a dialysis system in which high molecular weight growth factors could be retained while less costly low molecular weight nutrients (e.g. glucose, glutamine, amino acids, vitamins and buffers) could be added and waste products such as lactate and ammonia could be removed.

Initially we looked at lymphocytes as circulating blood cells and drew the analogy with B-cells and hybridomas. Earlier success with hybridomas in suspension culture led us to believe that it would be possible to cultivate cells at relatively high density if a new approach to oxygenation was used in conjunction with perfusion of the cells with fresh media. Reuveny (S. Reuveny, *et.al.*, *J. Immunol. Methods*, 86:53, 1986) has shown that hybridomas can be cultivated to densities of 5×10^7 and their goal was high antibody titer, not production of the cells themselves. These cells were maintained at high viabilities for weeks with reduced serum in the perfusing medium. Using one of several alternatives to increasing oxygen transport while avoiding fluid shear (caged aerators, increased partial pressure, external or internal membrane aerators of silicon or macroporous, hydrophobic polypropylene and/or stabilized hemoglobin) and perfusing with media containing high serum or other growth factor sources, we postulated that cells could be cultured at 10^8 cells/ml or above.

Further, if conservation of endogenously produced growth factors proved important, a dialysis device could be used to add low molecular weight nutrients and remove wastes. Such a bioreactor system is diagramed in figure 5.

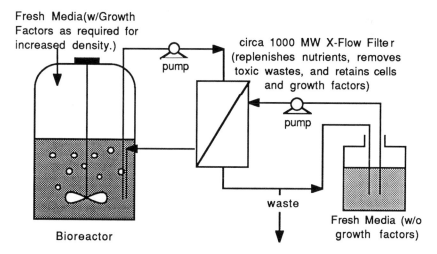

Figure 5. A suspension type bioreactor (essentially a spinner flask with added capabilities for control of pH, oxygen introduction, media addition and other factors not detailed. The filter represents several alternative devices or combinations of devices for retaining cells and growth factors while allowing nutrients and wastes to exchange with fresh media.

As noted above, TIL cells do not expand and appear to rapidly lose their activation in suspension culture. To date we have not evaluated the reactor described above using a very high inoculation density. If the "cells/surface area" constraint is accounted for solely by a requirement for endogenously generated growth factors (and not direct cell-cell interaction) then such a design may be satisfactory for the final stages of TIL expansion, e.g.from 10^9 to 10^{11} cells. Other approaches would be used to prepare the inoculum such as the current T-flask procedures.

T- Lymphocyte Proliferation In Vivo

Nature's bioreactors for the expansion of mature lymphocytes, the lymph nodes, are more similar to perfusion (high density) bioreactors than to suspension (relatively low density, high shear) bioreactors. The nodes consist of packed cells, essentially all lymphocytes and macrophages, with nutrients and antigen provided by blood and afferent lymphatic vessels. The fluids in these vessels essentially percolate through the dense cell mass. In pathological conditions, the nodes become the sites of formation of granular leukocytes, which may include the NK, LAK and activated T-lymphocytes similar to the populations we are cultivating in-vitro. In addition, the lymph nodes are thought to be the "traffic-control centers" which direct the migration of circulating lymphocytes in the body. At these high densities, both cell-cell interactions and high local concentrations of T-cell produced growth factors are possible.

Alternative Bioreactors

To reduce manipulation of tens of T-flasks, both Baxter Travenol and Dupont have introduced modified blood bags as T-flask substitutes. These may allow the required cells to be cultivated in fewer containers and with the use of gas permeable materials may resolve oxygen transfer constraints at moderate cell densities (circa 10^7 cells/ml) and relatively slow growth rates (doubling times of > 2 days). It is important to measure dissolved oxygen when such systems are introduced. This month, A. Nahapetian of Dupont reported on TIL cultivation in these bags at a symposium sponsored by the Northeast section of the American Society for Microbiology. A 10 fold reduction in media volume requirement as compared to Rosenberg's procedure was reported. Expansion of TIL derived using Rosenberg's procedure from 10^7 to 10^{10} cells was achieved in 30-40 days. This is a relatively slow growth rate (3 to 4 day doublings) compared to some we have achieved (2 day doubling times, see figures 4 and 5 above. Under improved nutritional conditions with recently reactivated TIL preparations we see doublings in the 24 hour range.

We are working with other bioreactor designs which may meet the practical and physiological requirements of lymphocyte expansion for LAK, TIL and other tissue replacement therapies. These are high density bioreactors which would simply replace the suspension reactor in figure 6 above. In the perfusion system, oxygenation requirements, etc. would remain the same. One of these reactors is a multi layer membrane device (Dynacell , Millipore) in which the cells are sandwiched between two media perfusion layers from which they are separated by a membrane (macroporous PVDF in hydrophilic form) which passes both nutrients, wastes and growth factors. A dialysis device would be separately required for retaining growth factors and thereby conserving the expensive portions of the media and allowing more rapid growth at lower cell density. When this device is used for hybridoma cultivation, approximately 5×10^9 viable cells can be recovered from it. Scale-up of the device to handle a full TIL or LAK dose should be straightforward.

In this type reactor, high cell densities and straightforward harvesting appears feasible with certain minor modifications. We have proposed a modification in which a third flow path would be added which would be separated from either the cell layer or the media layer by a membrane having a low MW cutoff. This path would only need to exchange with a fraction of the media path layers as the transport needs for exchanging low MW nutrients and wastes are far less than those for oxygen, which controls the design of the cell to media paths.

We are also examining other bioreactors, including one using a ceramic macroreticular matrix for immobilizing cells at high density. This matrix is used in conjunction with the Charles River Laboratory Opticell bioreactor. This bioreactor provides for monitoring and computer control of pH, oxygen consumption rate, and temperature, and includes automated seeding, feeding and harvesting. In work with other cells, high cell densities have been achieved in this system, as the cells adhere to and grow on the porous ceramic in an evenly distributed manner. We are thus investigating the growth of TILs in the Opticell. Modifications of the Opticell system were done in our laboratory to make it more suitable for this application. In order to decrease the adsorption of IL-2 to the ceramic, the core is soaked in a protein solution prior to use. Also, the size of the core was scaled down to reduce the required cell inoculum. We are currently looking at the recovery of the cells grown on the ceramic surface.

Control of Population Dynamics

Other aspects of control of culture conditions have to do with their influence on population dynamics. Both LAK and TIL procedures involve mixed populations of cells some of which, acting alone or in concert with other subpopulations, are responsible for the beneficial responses. Ultimately the selection of bioreactor type, media and process type will allow some control of cell physiology and perhaps of population dynamics so that the right combination of cells can be prepared. We are working to characterize individual T-cell classes in order to ascertain how to differentially stimulate or retard growth of subpopulations.

SUMMARY

"Adoptive" therapies must be seen in the context of therapies using lymphokines and MABs alone and in combination and with chemotherapy,etc. Today we do not know which combinations are desirable, nor do we know how the clinical response is mediated. We can anticipate that all of the issues discussed above will be important in optimizing the delivery of these therapies once more is known about the mechanisms of therapeutic responses. The systems described are quite flexible and will lend themselves to the use of various lymphokine and cytokine mixtures to control the physiology and differentiation of the cells for therapeutic use. Furthermore, systems of this type should be useful for preparation of other mammalian cells for therapy.

Tissue Engineering, pages 313–318
© 1988 Alan R. Liss, Inc.

ENGINEERING OF HEMATOPOIETIC CELLS

Jerome S. Schultz

Center for Biotechnology and Bioengineering
University of Pittsburgh
Pittsburgh, PA 15260

Many of the papers at this meeting consider procedures for the growth of tissues to provide structural function in the body, such as grafts for bone, connective tissue, and soft tissue. However, there are a great number disease states or impaired conditions where the need is to replace a biological or biochemical function rather than a structural function. In these cases the tissue need not be a continuous contiguous tissue but it actually can be a suspension of cells that can carry out specific biochemical or immunological functions. One such tissue that has the potential for multiple uses is the hematopoietic system. This tissue, where a multitude of different functional cell types are produced from a parent stem cell (which resides in the marrow) is a very rich system which can be utilized to replace many different biological functions.

Historically, blood or its various components such as red cells or platelets, or combinations thereof have been used as a liquid tissue supplement for patients who suffer from either an oxygen transfer deficit or bleeding. This rather ancient and well established technique of using blood components as a tissue replacement is now moving into a very advanced and modern capability of providing cells with many different and specifically designed functions. The advent of the discovery and ability to produce cell growth factors, along with the application of molecular biology for the modification of hematopoietic cells has opened a new era for the design and manipulation of these cell lines.

The capacity of the hematopoietic system to produce large quantities of cells is quite unusual. For example, in the normal person, each hour about ten billion red blood cells come to the end of their life span and are replaced by new red cells which derive from the stem cells. Also, the distribution of cells types produced from the stem cell is

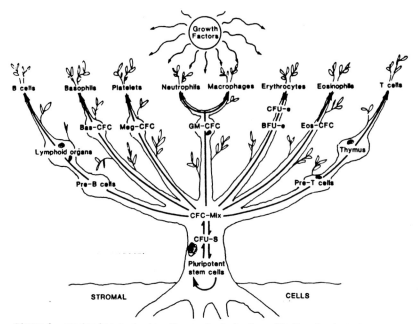

Figure 1. Murine haemopoiesis. Haemopoiesis begins with the pluripotent haemopoietic stem cell. This cell can either self-renew or undergo commitment to develop or differentiate into committed progenitor cells. The CFU-S (colony-forming unit-spleen) assay detects cells which belong to the stem cell compartment but CFU-S may not represent the total stem cell population. CFC-Mix (colony-forming cells-mixed) are cells which in semi-solid media can form multi-lineage colonies, and as such may have some overlap with those cells known as CFU-S. Committed progenitor cells of several distinct haemopoietic lineages have been identified in vitro, these include the basophil colony-forming cell (Bas-CFC); the megakaryocyte-CFC (Meg-CFC); the granulocyte/macrophage-CFC (GM-CFC); the burst-forming unit-erythroid (BFU-e); the more mature erythroid progenitor cell; the colony forming unit-erythroid (CFUI-E): and the eosinophil colony forming cell (Eos-CFC). Also, Pre-B lymphocytic cells and Pre-T lymphocytic cells can be recognized in appropriate assays in vitro. These cells are named colony-forming cells because of their ability to form a clonal colony of mature cells of a given lineage in semi-solid culture medium in vitro.

under fine control and is modulated according to the needs of the organism at any particular time, for example, under the stress of infection or bleeding. An unusual, but very important attribute of stem cells is that they are pluripotent, that is they have the genetic potential to undergo development to a variety of cell types. The determination of which cell type develops from a given stem cell depends in part on soluble growth factors (cytokines, lymphokines), there also appears to be a stochastic element as well. In recent years these growth factors have been isolated and cloned by recombinant DNA procedures so that now it is possible to develop specific cell lines from a given subpopulation by the appropriate utilization of growth factors in the culture media.

The types of cells that can be produced from a given stem cell (in the mouse) is shown in Figure 1 (Clark and Kamen, 1987), along with the known growth factors which regulate the formation of the various cell lines. Similar control mechanisms are being worked out for other mammalian systems as well. Each of these terminal cell types carries out a specific functions and is morphologically and genetically much different than the other product cells.

In recent years there have been some major clinical achievements made possible by the use of cells from the hematopoietic system to treat certain life threatening clinical conditions. Two current main applications of this liquid cell tissue are in the treatment of leukemias and aplasias by bone marrow transplantation, and the treatment of cancer by the reinfusion of activated lymphocyte killer cells.

In bone marrow transplantion it is necessary to insure that the recipient's immune system does not interfere with the newly transplanted cells. This is usually accomplished by ionizing radiation, which completely destroys the host's hematopoietic system. When the new stem cell population is introduced by injection, the new cells proliferate and then provide a whole new hematopoietic system to the recipient, which includes immunologically competent cells but also the ability to make erythrocytes, platelets and macrophages to manage the other functions of blood. The demonstration that a small sample of cells can be injected into a person and replace the entire cell population in blood over a period of several weeks demonstrates the potential value and the immense proliferative capacity of a small number of stem cells.

Figure 2 (Bortin and Rimm, 1986) shows the increase in the number of bone marrow cell transplants in recent years. It is clear that this type of tissue therapy is taking off exponentially. In some cases it has been observed that while the bone marrow as a therapy for a specific disease like leukemia, sometimes the transplanted cells also cured another disease of that individual because these new cells had a more normal complement of genes. A listing some of the diseases that can be treated by transplantation of bone marrow cells from one individual to another is given in Table 1 (Parkman, 1986).

There is another approach currently developing for the utilization of bone marrow cells as therapeutic agents. In this new emerging therapy stem cells would be removed from an individual and cultured in vitro so as to expand the population, then new genes would be added to these cells by

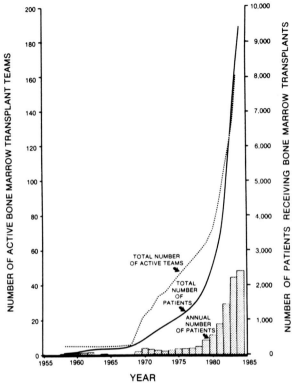

FIGURE 2. Vertical bars represent number of patients who received bone marrow transplants annually. Solid line indicates cumulative number of patients who received bone marrow transplants from 1958 through 1984. Dashed line represents cumulative number of active bone marrow transplant teams from 1958 through 1984.

Table 1. Classification of some of the genetic diseases that may be treated by bone marrow transplantation (BMT).

Class I	Class II	Class III	Class IV	Class V	Class VI
			Expression		
Expression of genetic defect restricted to lymphoid cells	Expression of genetic defect restricted to lymphoid and hematopoietic cells	Expression of genetic defect restricted to hematopoietic cells	Generalized genetic defect; clinical symptomatology restricted to lympho-hematopoietic cells	Generalized genetic defect; generalized clinical symptomatology with or without CNS involvement	Lympho-hematopoietic cells do not express the normal gene product
			Treatment		
Correctable by BMT	Correctable by BMT	Correctable by BMT	Correctable by BMT	May be correctable by BMT	Not correctable by BMT
			Disease		
Severe combined immune deficiency (non-ADA deficient)*	Wiskott-Aldrich syndrome*	Thalassemia*	Gaucher's disease*†	Adrenoleukodystrophy*	Cystic fibrosis
gpL-115 deficiency*	Chédiak-Higashi syndrome*	Granulocyte actin deficiency*	Adenosine deaminase deficiency*†	Metachromatic leukodystrophy*	Hemophilia
X-linked agamma-globulinemia		Chronic granulocyte disease*	Nucleotide phosphorylase deficiency*†	Krabbe's disease	Phenylketonuria
		Infantile agranulocytosis*	Fanconi anemia*	Mucopolysaccharidosis*†	
		Sickle cell disease*		Lesch-Nyhan syndrome*†	
		Osteopetrosis*			

*Diseases for which allogeneic bone marrow transplantation has been attempted.   †Candidate diseases for gene therapy.

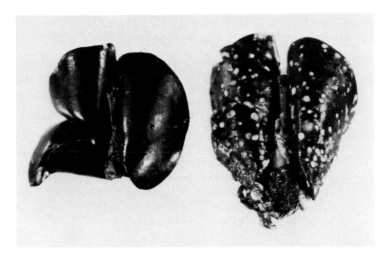

Figure 3. Photographs of lungs from F344 rats bearing 3-day established NK-resistant MADB106 mammary tumor and treated with 10^7 A-LAK cells plus rIL-2 (50×10^3 units, I.P. BID) left or rIL-2 alone (right). Lungs were removed on day 18.

---

means of recombinant DNA technologies. Now the new genetically modified cells with the new genes can be administered to the same person and thereby provide a cell population programmed to treat particular diseases. Since the patient receives their own cells, one would expect that these cells would be transparent to the recipient's immune system.

Recently Richard Mulligen (Sherr, 1987) showed that the stem cells can be genetically modified by the utilization of retroviral vectors. These transformed cells then can be used as therapeutic agents to treat diseases which could be helped by the production of specific proteins, like hormones, by this engineered cell system. Other possibilities are that these cells could be designed to produce some enzymes to overcome metabolic deficiencies.

The second current example is the use of lymphocyte activated killer cells for the treatment of cancer. In this procedure, samples of blood are taken from the patient and then the special killer cells are grown and the population increased by adding the chemical IL-2 to the system. The use of this growth factor allows these cells to multiply many fold and then subsequently after culturing for some time, the cells are reinfused into the same patient resulting in rather dramatic results, as shown in Figure 3 (Hiserodt, 1988).

This is but one example of many possible different cell therapies which could be based on any of the many different cell types that make up the hematopoietic system.

The efficient use of cells such as the killer lymphocytes for the treatment of cancer requires some new techniques to be developed for the isolation of these cells and their growth in culturing vessels. For example recently researchers at the University of Pittsburgh have shown that the biological activity of the cell preparation can be improved significantly if the particular LAK subpopulation of cells is utilized. This was obtained by a separation technique (involving specific cell adherence to plastic surfaces in the presence of IL-2) which allows the large granular lymphocytes to be removed and separated from the others; and then these particular cells can be activated in vivo by IL-2 at a much lower concentration.

In order to use these new cell-based technologies clinically, and to insure that they will be available to most medical centers, a number of simpler methodologies need to be developed for storing hematopoietic stem cells, regulating their differentiation, and controlling their proliferation. Particularly it will be necessary to develop culturing conditions for these cell lines.

Fortunately, in recent years a number of growth factors have been isolated and (some) genetically cloned so that sufficient amounts of these critical materials will be available for culturing hematopoietic cells in vitro. However at this time this particular technique is very cumbersome and although it can be made to work in a research laboratory setting, implementation in an ordinary medical clinic would be difficult. Thus there is a need for some creative engineering combined with biotechnology to develop procedures for growing these cells under well-controlled conditions and in a manner that is easy to adapt to ordinary clinical environments.

REFERENCES

Bortin, M.M. and Rimm, A.A., Transplantation, 42: 229-234 (1986).

Clark, C.C. and Kamen, Science, 236: 1229-1237 (1987).

Hiserodt, J. Personal Communication (1988).

Parkman, R., Science, 232: 1373-1378 (1986).

Sherr, C.J., Cell, 48: 727-729 (1987).

Vujanovic, N., Herberman, R., Maghazachi, A., and Hiserodt, J., J. Exp. Med., 167: 15-29 (1988).

Tissue Engineering, pages 319–320
© 1988 Alan R. Liss, Inc.

Summary/Discussion

The Hematopoietic System

Christopher Squier

Jerome Schultz reviewed evidence for modifying and supplementing functions of the haemopoietic system using growth factors for blood cells. In general these factors are small proteins with a very short half lives *in vivo* and are effective in extremely low concentrations.

Examples of existing applications of this approach are bone marrow transplantation in which deficiencies of the haemopoietic and/or lymphoid cells can be remedied. There are presently 25-30 diseases amenable to such treatment. However, the techniques of genetic engineering offer the exciting possibility of introducing new genes into haemopoietic stem cells that could cure diseases other than those of the haemopoietic system, such as inborn errors of metabolism. A related approach involves treatment of tumors with activated cells. Lymphocytes isolated from blood and activated with interleukine 2 (IL-2) show much higher anti-tumor activity than unactivated cells.

The advantages of the above approaches include the ability to culture the cells *in vitro,* the diversity of cell types available, the possibility of modifying the cells genetically, and the absence of histocompatibility problems.

The disadvantages of the approach are the problems of identifying various specific cell types, the difficulties of storage or of culture in a clinical setting, the uncertain mode of action of growth factors and our ignorance about other factors. A major problem with using lymphoid cells is the separation of classes of T-lymphocytes, so that nonrandom mixtures may result. Similarly, there would be advantages in isolating specific haemopoietic progenitor cells, but identification is a problem.

In terms of culture techniques, a closed system roller bottle culture device was described, and it was agreed that there is a need for these techniques to be scaled up now that the construction is feasible.

Randy Swartz described the current state of the art for producing LAK and TIL cells. He described the deficiencies of current systems in terms of physiological environment, the need for better definition of specific cell sub-populations, and the need for more efficient processing systems (bioreactors, etc.) to effectively reduce cost while at the same time producing more defined cell populations for therapeutic uses.

Tissue Engineering, pages 321–323
© 1988 Alan R. Liss, Inc.

VIII. General Discussion: Mathematical Modeling

Richard Skalak

As the processes involved in tissue engineering become more complex and quantitative data on the results become available, there is a need and opportunity in the construction of mathematical models. Such modeling becomes essential to check whether hypotheses concerning the important factors involved in a process are physically feasible and to suggest how to optimize the results. A principal area of current need in modeling is to describe the effect of stress on growth, both of cells and tissues. However, the influence of electrical fields also needs fundamental experiments and models to demonstrate basic biophysical effects involved. Without modeling, experiments with electrical fields may remain quite mysterious as to their manner of producing observed results.

Mathematical modeling is a good example of one aspect of the interdisciplinary nature of tissue engineering. The stress analysis, electrical field theory, mathematical formulation and computational techniques required are becoming more advanced and require professional personnel in expertise in these areas for their efficient accomplishment. As the availability and complexity of mathematical, computer and graphic systems become more advanced, it is requisite for efficiency to have people with professional background to develop these aspects.

In application of mathematical modeling to biological problems in general, probably the two most highly developed areas are in wave propagation in blood flow and in mechanical analysis within orthopedics applications. However, most of the literature deals with macroscopic and passive effects. The literature concerning growth of individual cells is just now developing and represents an opportunity and a need in respect to tissue engineering. Detailed evidence of the effect of stress fields on individual cells is beginning to accumulate and needs fundamental, mathematical modeling at the internal cellular level for explanation. An example of the influence of mechanical stresses on cellular behavior is given by R.M. Nerem describing endothelial cell responses to fluid shear stresses. This includes decreased cell proliferation, changes of shape, orientation and endocytosis. These effects may be useful in designing optimal vascular grafts, and mathematical models of a cellular behavior are needed for such tasks.

Several papers presented at the Workshop discussed stress induced remodeling in the musculoskeletal system. An important area is the influence of stress and strain on cells grown in a reconstituted matrix, as discussed by H. Alexander for osteoblasts in a collagen matrix. Somewhat different theories have been presented by D. Carter and S.E. Cowin on quantitative stress growth laws for natural development and due to imposed stress field. It is a challenge to bring such theories into a firm agreement with experimental data and then to see if such effects can be used to optimize the development of tissues for tissue engineering.

A dramatic and cogent aspect of stresses in tissues is brought out by Professor Y.C. Fung. The key concept is the role of residual stress, which tends in general to produce a more uniform distribution under normal loads. This is demonstrable in the arterial walls and the heart. It remains to be seen if residual stress effects can be deliberately built into artificial tissues in a beneficial way.

A broad challenge to mathematical modeling in tissue engineering is to optimize the combined effects of mechanical stress, biochemical and electrical fields applied to cells and tissues to produce arrangements and forms desired for specific tissue applications. Effects of electrical stimulations of musculoskeletal growth and remodeling is discussed by J. Black. It is clear that electrical stimulation affects various aspects of cellular behavior, as much as biochemical and mechanical conditions. The need is now to develop comprehensive theories of the interactive effects of mechanical, electrical and chemical environments imposed simultaneously on individual cells as well as complete tissues.

In summary, mathematical and computational models are needed in relation to tissue engineering for:

(1) Fundamental understanding of the physical processes involved in cell division, adhesion, assembly in growth or hypertrophy of tissues.

(2) Testing quantitative formulation of biological processes such as the influence of mechanical stresses and electrical fields on growth and form of

tissues.

(3) Allowing prediction of results of proposed experiments or new treatment modalities in advance of performing experiments for efficiency.

(4) Optimizing development and form and properties of tissues for tissue engineering by modeling electrical, chemical and mechanical fields in a forward and logical way.

Acknowledgments

The editors gratefully acknowledge the following individuals who contributed to discussions in this volume:

Fred Heineken   K.R. Sprugel   V.B. Hatcher
R.M. Nerem   L.M. Bjursten   S.L.Y. Woo
S.C. Cowin   Van C. Mow   M. Bernfield
R.W. Swartz   D.D. Cunningham   P. Richardson
H. Alexander   Y.C. Fung   D. Carter

Index